T0202567

SpringerBriefs in Optimization

SpringerBriefs present concise summaries of cutting-edge research and practical applications across a wide spectrum of fields. Featuring compact volumes of 50 to 125 pages, the series covers a range of content from professional to academic. Briefs are characterized by fast, global electronic dissemination, standard publishing contracts, standardized manuscript preparation and formatting guidelines, and expedited production schedules.

Typical topics might include:

- A timely report of state-of-the art techniques
- A bridge between new research results, as published in journal articles, and a contextual literature review
- A snapshot of a hot or emerging topic
- An in-depth case study
- A presentation of core concepts that students must understand in order to make independent contributions

SpringerBriefs in Optimization showcase algorithmic and theoretical techniques, case studies, and applications within the broad-based field of optimization. Manuscripts related to the ever-growing applications of optimization in applied mathematics, engineering, medicine, economics, and other applied sciences are encouraged.

Titles from this series are indexed by Web of Science, Mathematical Reviews, and zbMATH.

More information about this series at https://link.springer.com/bookseries/8918

Qiao-Li Dong • Yeol Je Cho • Songnian He
Panos M. Pardalos • Themistocles M. Rassias

The Krasnosel'skiĭ–Mann Iterative Method

Recent Progress and Applications

 Springer

Qiao-Li Dong
College of Science
Civil Aviation University of China
Tianjin, China

Songnian He
College of Science
Civil Aviation University of China
Tianjin, China

Themistocles M. Rassias
Department of Mathematics
National Technical University of Athens
Athens, Greece

Yeol Je Cho
Department of Mathematics Education
Gyeongsang National University
Jinju, Korea (Republic of)

Panos M. Pardalos (iD)
Department of Industrial & Systems Eng.
University of Florida
Gainesville, FL, USA

ISSN 2190-8354 ISSN 2191-575X (electronic)
SpringerBriefs in Optimization
ISBN 978-3-030-91653-4 ISBN 978-3-030-91654-1 (eBook)
https://doi.org/10.1007/978-3-030-91654-1

Mathematics Subject Classification: 26-XX, 28-XX, 35-XX, 39-XX, 41-XX, 46-XX, 47-XX, 49-XX, 58-XX, 65-XX, 68-XX, 90-XX, 91-XX, 93-XX

This Springer imprint is published by the registered company Springer Nature Switzerland AG
The registered company address is: Gewerbestrasse 11, 6330 Cham, Switzerland

Preface

The Krasnosel'skiĭ–Mann (KM) iterative method has extensively been employed to find fixed points of nonlinear mappings (in particular, nonexpansive mappings) and solve convex optimization problems. It is well known that several optimization algorithms are indeed special cases of the Krasnosel'skiĭ–Mann iteration such as the forward-backward splitting algorithm, the backward-forward splitting algorithm, the Douglas–Rachford splitting algorithm, and the Davis–Yin splitting method. In this book, first we present the development of the original Krasnosel'skiĭ–Mann iteration and discuss the convergence rate of the Krasnosel'skiĭ–Mann iteration and its perturbations. Second, we focus on the inertial Krasnosel'skiĭ–Mann (iKM) iteration and investigate the choice of the inertial parameters. Further, we discuss the multi-step inertial Krasnosel'skiĭ–Mann (MiKM) iteration and analyze its convergence and the choice of the inertial parameters. Third, we discuss the selections of relaxation parameters of the Krasnosel'skiĭ–Mann iteration and, especially, present the approximation for the optimal relaxation parameter sequence. Also, we introduce some iterative schemes based on the projection methods for variational inequality problems and the residue algorithm for the nonlinear monotone equations, which are shown to be the (multi-step) inertial Krasnosel'skiĭ–Mann iteration. Finally, we present two applications of the Krasnosel'skiĭ–Mann iteration in large scale optimization problems such the asynchronous parallel coordinate updates methods and cyclic coordinate update algorithms. We also suggest several open problems for further research in these areas.

Tianjin, China   Qiao-Li Dong
Tianjin, China   Songnian He
Jinju, Republic of Korea   Yeol Je Cho
Gainesville, FL, USA   Panos M. Pardalos
Athens, Greece   Themistocles M. Rassias

Contents

Chapter 1
Introduction

Let $\mathcal{H}$ be a Hilbert space, C a nonempty closed convex subset of $\mathcal{H}$, and $T : C \to C$ an operator. Denote by $\mathrm{Fix}(T)$ the set of fixed points of T, i.e., $\mathrm{Fix}(T) = \{x \in C : x = T(x)\}$. The *fixed point problem* can be formulated as follows:

$$\text{Find } x^* \in C \text{ such that } T(x^*) = x^*. \tag{1.1}$$

The fixed point problem (1.1) has wide applications in dynamic systems theory, integral and differential equations and inclusions, mathematics of fractals, mathematical economic (game theory and equilibrium problem) [150], mathematical modelling and optimization problems.

In connection with the fixed point problem, the following questions were investigated:

(a) Does a fixed point exist?
(b) Do we have the uniqueness of the fixed point?
(c) When a fixed point exists, how do we approximate it?

Let $T : \mathcal{H} \to \mathcal{H}$ be an operator. If T is contractive, then it has a unique fixed point by Banach contraction principle. However, if T is merely nonexpansive, the situation is quite different. Indeed, a nonexpansive operator may have no fixed point (take $T : x \mapsto x + z$, with $z \neq 0$), exactly one (take $T = -\mathrm{Id}$, "Id" denotes the identity mapping on $\mathcal{H}$), or infinitely many (take $T = \mathrm{Id}$). Even firmly nonexpansive operators can fail to have fixed points (see [47, Example 22]).

In this paper we focus on (c), i.e., approximating a fixed point of a nonlinear operator provided that its fixed point exists. The constructive methods used in fixed point theory are prevailingly iterative procedures, that is, successive approximate methods. A great deal of iterative methods are designed to solve the fixed point problem (1.1) and it is difficult to list all of them. Here we briefly introduce some classical methods. Interested readers may consult the monographs by Berinde [18] and Chidume [37].

Q.-L. Dong et al., *The Krasnosel'skiĭ-Mann Iterative Method*, SpringerBriefs in Optimization, https://doi.org/10.1007/978-3-030-91654-1_1

1.1   Fixed Point Iteration Procedures

In this section, we present some classical fixed point iteration procedures in a Hilbert space $\mathscr{H}$.

(A) The Picard iteration

Let C be a nonempty closed convex set of $\mathscr{H}$ and $T : C \to C$ be a mapping. For an arbitrary $x_0 \in C$, the *Picard iteration* $\{x_n\}$ is defined as follows:

$$x_{n+1} = T(x_n) \quad \text{for each} \quad n \geq 0. \tag{1.2}$$

The Picard iteration introduced in [171] is the simplest fixed point iteration procedure and converges when $T : C \to C$ is a contractive mapping (see its definition in Chap. 2). By using the Banach contraction mapping principle, one has the priori and posteriori error estimations of the Picard iteration, from which the stopping criterion is offered.

The Picard iteration may fail to produce a fixed point of $T : C \to C$ when T is nonexpansive. A simple illustration of this situation is $T = -\text{Id}$ and $x_0 \neq 0$.

(B) The Krasnosel'skiĭ iteration

To construct the convergent sequence for nonexpansive mappings, the relaxation variants of the Picard iteration were introduced. Krasnosel'skiĭ [120] firstly presented the following midpoint averaged method involving two successive terms of the Picard iteration:

$$x_{n+1} = \frac{1}{2}x_n + \frac{1}{2}T(x_n) \quad \text{for each} \quad n \geq 0, \tag{1.3}$$

where $x_0 \in C$ and $T : C \to C$ is a mapping.

Krasnosel'skiĭ presented the convergence result in a Banach space:

Theorem 1.1 ([120, Theorem 1]) *Let X be a uniformly convex Banach space and C be a convex compact subset of X. If $T : C \to C$ is a nonexpansive mapping with $\text{Fix}(T) \neq \emptyset$, then $\{x_n\}$ defined by (1.3) converges strongly to a fixed point of T.*

Edelstein [79] weakened the uniform convexity to the strict convexity of the above theorem. Schaefer [179] extended Krasnosel'skiĭ's result to a more general case by replacing the $\frac{1}{2}$ in (1.3) with the constant $\lambda \in (0, 1)$:

$$x_{n+1} = (1 - \lambda)x_n + \lambda T(x_n) \quad \text{for each} \quad n \geq 0. \tag{1.4}$$

The weak convergence of the Krasnosel'skiĭ iteration (1.4) was first proved by Schaefer [179] for a class of continuous nonexpansive operators. It is easy to see that the iteration (1.4) is exactly the Picard iteration corresponding to the averaged operator

$$T_\lambda = (1 - \lambda)\text{Id} + \lambda T$$

and, for $\lambda = 1$, the iteration (1.4) reduces to the Picard iteration. Moreover, we have Fix$(T) = $ Fix(T_λ) for all $\lambda \in (0, 1]$.

The most general Krasnosel'skiĭ iteration was given by replacing λ with $\lambda_n \in [0, 1]$ in the iteration (1.4):

$$x_{n+1} = (1 - \lambda_n)x_n + \lambda_n T(x_n) \quad \text{for each } n \geq 0. \tag{1.5}$$

This iteration has been investigated in a great deal of papers in the last sixty years and is well known as the *Krasnosel'skiĭ–Mann iteration*. We will discuss in detail it in Chap. 3.

(C) The Ishikawa iteration

For an arbitrary $x_0 \in C$, the *Ishikawa iteration* defines an iterative sequence $\{x_n\}$ as follows: for each $n \geq 0$,

$$\begin{cases} y_n = (1 - \alpha_n)x_n + \alpha_n T(x_n), \\ x_{n+1} = (1 - \lambda_n)x_n + \lambda_n T(y_n). \end{cases} \tag{1.6}$$

The Ishikawa iteration (1.6) named after its inventor [111] was first used to establish the strong convergence to a fixed point for a Lipschitz and pseudocontractive self-mapping of a convex compact subset C of a Hilbert space $\mathcal{H}$.

Chidume and Mutangadura [38] constructed an example of a Lipschitz pseudocontractive mapping with a unique fixed point for which the sequence generated by the Krasnosel'skiĭ–Mann iteration (1.5) fails to converge. Therefore, the Ishikawa iteration is generally used to approximate fixed points of Lipschitz pseudocontractive operators. Interest in pseudocontractive mappings stems mainly from their firm connection with the class of nonlinear accretive operators. It is a classical result that, if T is an accretive operator, then the solutions of the equation $T(x) = 0$ correspond to the equilibrium points of some evolution systems [56].

The Ishikawa iteration (1.6), which is actually the two-step Krasnosel'skiĭ–Mann iteration, is computationally more complicated than the Krasnosel'skiĭ–Mann iteration. Furthermore, the convergence rate of the Ishikawa iteration is slower than that of the Krasnosel'skiĭ–Mann iteration [63]. From a practical point of view, when two or more fixed point iterative schemes are known to be convergent in a certain class of mappings, it is natural to choose the simplest method among them.

In infinite dimensional spaces, the convergence of the Krasnosel'skiĭ–Mann iteration and the Ishikawa iteration is generally weak. Genel and Lindenstraus [85] gave a celebrated counterexample of a contractive operator defined on a bounded closed convex subset C of a Hilbert space $\mathcal{H}$ for which the Krasnosel'skiĭ iteration does not strongly converge. Another counterexample can be found in [109] as for iterations of the composition of two projections, i.e., the Picard iteration for averaged operator. However, the strong convergence is often much more desirable than the weak convergence in many problems that arise in infinite dimensional spaces (see [16] and references therein). Thus additional conditions on the operators or some modifications of the Krasnosel'skiĭ–Mann iteration (1.5) are necessary to get the strong convergence.

First, we discuss additional conditions on the operators. If T is demicontractive and Id $- \; T$ maps a closed bounded subset of C onto a closed subset of C (in particular, if T is demicompact), then the Krasnosel'skiĭ–Mann iteration (1.5) converges strongly to a fixed point of T (see [104]).

Maruster [146] gave another condition that there exists $h \in C$ such that $h \neq 0$ and

$$\langle x - T(x), h \rangle \leq 0 \quad \text{for all} \;\; x \in C.$$

The point h satisfying the above condition was defined for linear equations in Hilbert spaces [146], which is generally difficult to obtain.

Next, we consider some modifications of the Krasnosel'skiĭ–Mann iteration. There are two types of modifications with the strong convergence, that is, the Halpern iteration (viscosity iteration) and the hybrid methods.

(D) The Halpern iteration

For any fixed $u \in C$ and arbitrary $x_0 \in C$, the Halpern iteration [90] is defined as follows:

$$x_{n+1} = \lambda_n u + (1 - \lambda_n) T(x_n) \quad \text{for each} \;\; n \geq 0, \tag{1.7}$$

in which $\{\lambda_n\} \subset [0, 1]$ and the convex combination used to define x_{n+1} is "anchored" to u.

Let $T : C \to C$ be a nonexpansive mapping with Fix$(T) \neq \emptyset$. The parameter sequence $\{\lambda_n\}$ is assumed to satisfy the following conditions:

(C1) $\lim_{n \to \infty} \lambda_n = 0$;
(C2) $\sum_{n=0}^{\infty} \lambda_n = \infty$;
(C3) $\sum_{n=0}^{\infty} |\lambda_{n+1} - \lambda_n| < \infty$ or $\lim_{n \to \infty} \frac{\lambda_n}{\lambda_{n+1}} = 1$.

Then the sequence $\{x_n\}$ generated by the Halpern iteration (1.7) strongly converges to $P_{\text{Fix}(T)} u$, where P_C denotes the metric projection onto the set $C \subseteq \mathcal{H}$ (see, for example, [200]). Very recently, the tight convergence rate for the Halpern iteration with $\lambda_n = \frac{1}{n+2}$ and $u = x_0$ is given in [128] as follows:

$$\|x_n - T(x_n)\| \leq \frac{2\|x_0 - x^*\|}{n + 1} \quad \text{for each} \;\; n \geq 1 \;\; \text{and} \;\; x^* \in \text{Fix}(T).$$

In fact, the convergence speed of the Halpern iteration is widely believed to be slow due to the strict restriction (C2) on the parameter sequence $\{\lambda_n\}$ (see [145]).

In [95], He et al. introduced two optimal choices of the parameters λ_n, one of which is defined by

$$\lambda_n = \begin{cases} 0, & \text{if } u - 2T(x_n) + T^2(x_n) = 0, \\[2ex] \dfrac{\|T(x_n) - T^2(x_n)\|^2 - \|u - T(x_n)\|^2}{2\|u - 2T(x_n) + T^2(x_n)\|^2} + \dfrac{1}{2}, & \text{if } u - 2T(x_n) + T^2(x_n) \neq 0. \end{cases}$$

$$(1.8)$$

The Halpern iteration with $\{\lambda_n\}$ in (1.8) highly improves the convergence speed of that with $\{\lambda_n\}$ satisfying (C1)–(C3) (see numerical examples in [95]).

The minimum norm fixed point of a nonexpansive mapping is generally desirable since it is closely related to convex optimization problems. In fact, one can obtain the minimum norm fixed point of T if u in (1.7) is taken as 0. However, if $0 \notin C$, then the Halpern iteration process $\{x_n\}$ defined by

$$x_{n+1} = (1 - \lambda_n)T(x_n) \quad \text{for each } n \geq 0,$$

cannot be used for finding the minimum norm fixed point of T since $\{x_n\}$ may not belong to C. In order to overcome this weakness, Wang and Xu [194] introduced the following iteration process:

$$x_{n+1} = P_C[(1 - \lambda_n)T(x_n)], \tag{1.9}$$

where $\{\lambda_n\}$ is assumed to satisfy the conditions (C1)–(C3). However, it is difficult to implement the iteration process (1.9) in the actual calculations because the specific expression of P_C cannot be obtained, in general.

To avoid computing the projection onto C, He et al. [96] proposed an explicit boundary point algorithm. Define a function $h : C \to (0, 1]$ by

$$h(x) = \inf\{\lambda \in (0, 1] : \lambda x \in C\} \quad \text{for all } x \in C. \tag{1.10}$$

The explicit boundary point algorithm for finding the minimum norm fixed point of T is designed by

$$x_{n+1} = \lambda_n \theta_n x_n + (1 - \lambda_n)T(x_n) \quad \text{for each } n \geq 0, \tag{1.11}$$

where $\{\lambda_n\} \subset (0, 1)$, $\theta_n = h(x_n)$ and x_0 is taken in C arbitrarily.

The convergence of the explicit boundary point algorithm (1.11) is presented in the following theorem.

Theorem 1.2 *Let C be a closed convex subset of a Hilbert space $\mathscr{H}$ and $T : C \to C$ a nonexpansive mapping with $\mathrm{Fix}(T) \neq \emptyset$. Assume that the sequences $\{\lambda_n\}$ and $\{\theta_n\}$ of real numbers satisfy the following conditions:*

(C4) $\lambda_n \to 0$ *and* $\sum_{n=0}^{\infty} \lambda_n = \infty$;
(C5) $\limsup_{n \to \infty} \theta_n \leq \bar{\theta} < 1$;
(C6) $\sum_{n=0}^{\infty} |\lambda_n - \lambda_{n-1}| < \infty$ *or* $\lim_{n \to \infty} \frac{\lambda_n}{\lambda_{n-1}} = 1$;
(C7) $\sum_{n=0}^{\infty} \lambda_n |\theta_n - \theta_{n-1}| < \infty$ *or* $\lim_{n \to \infty} \frac{\theta_n}{\theta_{n-1}} = 1$.

Then the sequence $\{x_n\}$ generated by (1.11) converges strongly to $P_{Fix(T)}0$.

The main difficulty of using the boundary point algorithm (1.11) is to get the explicit expression of the function h defined in (1.10). A simple example was provided in [96]. Let $C = \{x \in \mathcal{H} | \varphi(x) \le 0\}$ where $\varphi : \mathcal{H} \to \mathbb{R}$ defined by

$$\varphi(x) = \|x - x_0\|^2 + \langle x, v \rangle \quad \text{for all } x \in \mathcal{H},$$

where x_0 and v are two given points in $\mathcal{H}$ such that $\langle x_0, v \rangle < 0$. Then $h(x)$ corresponding C is given by

$$h(x) = \frac{2\langle x, x_0 \rangle + \langle x, v \rangle - \sqrt{(\langle x, v \rangle - 2\langle x, x_0 \rangle)^2 + 4\|x\|^2 \|x_0\|^2}}{2\|x\|^2}.$$

(E) The viscosity iteration

Moudafi [154] extended the Halpern iteration and introduced the following viscosity iteration:

$$x_{n+1} = \lambda_n f(x_n) + (1 - \lambda_n)T(x_n) \quad \text{for each } n \ge 0, \tag{1.12}$$

where $f : C \to C$ is a contractive operator. The scheme (1.12) is called the viscosity iteration, since it represents a "brake" in the original Krasnosel'skiĭ–Mann iteration (1.5). The introduction of the viscosity term is not a simple formal object, but its significance is substantial.

Moudafi [154] firstly presented the convergence theorem of the viscosity iteration (1.12) and then Xu [199] relaxed the conditions on the parameter sequence $\{\lambda_n\}$ and proved the following result in a Hilbert space.

Theorem 1.3 ([199]) *Let $T : C \to C$ be a nonexpansive mapping with $Fix(T) \ne \emptyset$ and $f : C \to C$ be a contractive operator with a constant $\alpha \in [0, 1)$. Let $\{x_n\}$ be generated by (1.12). Then, under the hypotheses (C1)–(C3), the sequence $\{x_n\}$ converges strongly to a point in $Fix(T)$ which is the unique solution of the following variational inequality problem:*

$$\text{Find} \quad q \in Fix(T) \quad \text{such that} \quad \langle f(q) - q, x - q \rangle \le 0 \quad \text{for all } x \in Fix(T). \tag{1.13}$$

Xu [199] also extended the convergence of the viscosity iteration (1.12) to a uniformly smooth Banach space.

We can see that the mapping $Id - f$ is $(1 + \alpha)$-Lipschitz continuous and $(1 - \alpha)$-strongly monotone (see their definitions in Chap. 2). Therefore, the variational inequality problem (1.13) has a unique solution.

There are a lot of extensions and generalizations of the viscosity iteration (1.12) (see, for example, [61, 73, 133, 144, 191]). Similar to (1.8), He et al. [95] also presented the optimal selections of the parameter sequence $\{\lambda_n\}$ of the viscosity iteration (1.12).

Yao et al. [203] introduced the following modified Krasnosel'skiĭ–Mann iteration which is also called regularized Krasnosel'skiĭ–Mann iteration (see [136]): for each $n \geq 0$,

$$
\begin{cases}
y_n = \beta_n x_n, \\
x_{n+1} = (1 - \lambda_n) y_n + \lambda_n T(y_n),
\end{cases}
\tag{1.14}
$$

where $T : \mathscr{H} \to \mathscr{H}$ is a nonexpansive mapping. The strong convergence of the iterative sequence $\{x_n\}$ generated by (1.14) was shown provided that $\{\lambda_n\}$ and $\{\beta_n\}$ in $[0, 1]$ satisfy the following conditions:

(a) $\lim_{n \to \infty} \beta_n = 1$, (b) $\sum_{n=0}^{\infty} (1 - \beta_n) = \infty$, (c) $\lambda_n \in [a, b] \subset (0, 1)$.

Recently, Bot et al. [25] changed the convergence conditions in [203] and gave the following results:

Theorem 1.4 ([25, Theorem 3]) *Let $\{\lambda_n\}$ and $\{\beta_n\}$ be the real sequences satisfying the following conditions:*

(C9) $0 < \beta_n \leq 1$ *for each $n \geq 0$,*

$$
\lim_{n \to \infty} \beta_n = 1, \quad \sum_{n=0}^{\infty} (1 - \beta_n) = \infty, \quad \sum_{n=1}^{\infty} |\beta_n - \beta_{n-1}| < \infty;
$$

(C10) $0 < \lambda_n \leq 1$ *for each $n \geq 0$,*

$$
\liminf_{n \to \infty} \lambda_n > 0, \quad \sum_{n=1}^{\infty} |\lambda_n - \lambda_{n-1}| < +\infty.
$$

Let $T : \mathscr{H} \to \mathscr{H}$ be a nonexpansive mapping with $Fix(T) \neq \emptyset$ and $\{x_n\}$ be the sequence generated by the scheme (1.14). Then the sequence $\{x_n\}$ converges strongly to $P_{Fix(T)}0$.

(F) The hybrid methods

In his unpublished dissertation, Haugazeau [92] proposed independently a strongly convergent variant of the alternated projection method. Thereafter Bauschke and Combettes [16] extended Haugazeau's idea to some iterative methods in optimization problems. By combining the Krasnosel'skiĭ–Mann iteration and the Haugazeau algorithm, Nakajo and Takahashi [158] first introduced the following hybrid algorithm: for each $n \geq 0$,

$$\begin{cases} x_0 \in C \text{ chosen arbitrarily}, \\ y_n = \lambda_n x_n + (1 - \lambda_n) T(x_n), \\ C_n = \{z \in C : \|y_n - z\| \le \|x_n - z\|\}, \\ Q_n = \{z \in C : \langle x_n - z, x_n - x_0 \rangle \le 0\}, \\ x_{n+1} = P_{C_n \cap Q_n} x_0. \end{cases} \tag{1.15}$$

It is easy to see that the hybrid method (1.15) is a projection modification of the Krasnosel'skiĭ–Mann iteration. Its main idea consists in projecting the initial iterate x_0 onto the intersection of the suitably constructed sets C_n and Q_n that depend on the current iteration x_n. For this reason, the scheme (1.15) is often called the *CQ algorithm*.

Assume that $T : C \to C$ is a nonexpansive mapping with $\text{Fix}(T) \ne \emptyset$ and $\{\lambda_n\} \subset [0, \sigma]$ for some $\sigma \in [0, 1)$. Then the sequence $\{x_n\}$ generated by the hybrid method (1.15) strongly converges to $P_{\text{Fix}(T)}(x_0)$. It is obvious that the conditions of the strong convergence of the hybrid method (1.15) are weaker than those of the Halpern iteration (1.7).

Since the inception of the hybrid method (1.15), a great deal of modifications were proposed (see, for example, [62, 64, 123, 143, 182, 187, 206]), the most famous of which is the shrinking projection method [189] as follows: let $x_0 \in \mathscr{H}$, $C_1 = C$, $x_1 = P_{C_1} x_0$ and, for each $n \ge 0$,

$$\begin{cases} y_n = \lambda_n x_n + (1 - \lambda_n) T_n(x_n), \\ C_{n+1} = \{z \in C_n : \|y_n - z\| \le \|x_n - z\|\}, \\ x_{n+1} = P_{C_{n+1}} x_0. \end{cases} \tag{1.16}$$

Note that C_n in (1.16) is the intersection of finitely many closed sets and a half-space is added after each iterative step. This limits the range of the use of the scheme (1.16) (see, e.g., [107]).

Another modification introduced in [70] was paid much attention to by researchers due to its efficient numerical performance and its formula is addressed as follows: for each $n \ge 0$,

$$\begin{cases} x_0, z_0 \in C \text{ chosen arbitrarily}, \\ z_{n+1} = \lambda_n z_n + (1 - \lambda_n) T(x_n), \\ C_n = \{z \in C : \|z_{n+1} - z\|^2 \le \alpha_n \|z_n - z\|^2 + (1 - \alpha_n)\|x_n - z\|^2\}, \\ Q_n = \{z \in C : \langle x_n - z, x_n - x_0 \rangle \le 0\}, \\ x_{n+1} = P_{C_n \cap Q_n} x_0, \end{cases} \tag{1.17}$$

where $\{\lambda_n\} \subset [0, \sigma]$ for some $\sigma \in [0, \frac{1}{2})$. The convergence of the sequence $\{x_n\}$ is the same as that of (1.15). Note that $z_{n+1} = \lambda_n z_n + (1 - \lambda_n)T(x_n)$ in (1.17) can be replaced with $z_{n+1} = \lambda_n x_n + (1 - \lambda_n)T(z_n)$.

It was pointed out in [136] that the practical computation of the value $x_{n+1} = P_{C_n \cap Q_n} x_0$ seems to be a nontrivial task. Numerous papers have been published on this subject, investigating various aspects of the hybrid methods. Nevertheless, to our knowledge, there are very few papers concerning the implementation of the hybrid methods.

He et al. [98] first considered the realization of the hybrid method (1.15) and presented the explicit expression of the projections onto $C_n \cap Q_n$, when $C = \mathcal{H}$ as follows:

$$
\begin{cases}
x_0 \in \mathcal{H} \text{ chosen arbitrarily,} \\[4pt]
x_{n+1} = x_n, \text{ if } T(x_n) = x_n, \\[4pt]
y_n = \lambda_n x_n + (1 - \lambda_n)T(x_n), \text{ if } T(x_n) \neq x_n, \\[4pt]
u_n = \dfrac{1}{2}(x_n + y_n), \\[4pt]
C_n = \{z \in \mathcal{H} : \langle x_n - u_n, z - u_n \rangle \leq 0\}, \\[4pt]
Q_n = \{z \in \mathcal{H} : \langle x_0 - x_n, z - x_n \rangle \leq 0\}, \\[4pt]
x_{n+1} = u_0, \text{ if } n = 0, \\[4pt]
x_{n+1} = z_n, \text{ if } n \geq 1, \text{ and } z_n \in Q_n, \\[4pt]
x_{n+1} = w_n, \text{ if } n \geq 1, \text{ and } z_n \notin Q_n,
\end{cases}
\tag{1.18}
$$

where

$$
z_n = x_0 - \frac{\langle x_n - u_n, x_0 - u_n \rangle}{\|x_n - u_n\|^2}(x_n - u_n)
$$

and

$$
w_n = \left(1 - \frac{\langle x_0 - x_n, x_n - u_n \rangle}{\langle x_0 - x_n, z_n - u_n \rangle}\right) u_n + \frac{\langle x_0 - x_n, x_n - u_n \rangle}{\langle x_0 - x_n, z_n - u_n \rangle} z_n.
$$

When $C \neq \mathcal{H}$, the structure of $C_n \cap Q_n$ may be very complicated in general and so the calculation of $P_{C_n \cap Q_n}$ is difficult even if C has a perfect structure. To overcome this difficulty, recently, He and Yang [97] proposed the following modified successive projection method: for each $n \geq 0$,

$$\begin{cases} x_0 \in C \text{ chosen arbitrarily,} \\ y_n = \lambda_n x_n + (1 - \lambda_n) T(x_n), \\ C_n = \{z \in \mathcal{H} : \|y_n - z\| \leq \|x_n - z\|\}, \\ Q_n = \{z \in \mathcal{H} : \langle x_n - z, x_n - x_0 \rangle \leq 0\}, \\ u_{n+1} = P_{C_n \cap Q_n} x_0, \\ x_{n+1} = P_C u_{n+1}. \end{cases} \tag{1.19}$$

Note that, except for the strong convergence, another advantage of the Halpern iteration and the hybrid methods is that the sequence generated by them converges to a fixed point which one expects, such as a minimum norm fixed point. By using this property, the approximating methods were introduced to solve the split equality problem [68] and other nonlinear problems.

1.2   Fixed Point Formulation of Typical Problems

In this section, we discuss the fixed point formulation of nonlinear equations, integral equations and ordinary differential equations, and equilibrium problems.

(A) The nonlinear equation

Consider the following nonlinear equation:

$$F(x) = 0, \tag{1.20}$$

where $F : C \subseteq \mathbb{R}^n \to \mathbb{R}^n$ is a given operator. Efficiently finding roots of the nonlinear equation (1.20) is of major importance and has significant applications in numerical mathematics. In general, one could not directly solve the nonlinear equation (1.20) and needs to resort to the iterative methods.

One of the most used methods is to write (1.20) equivalently in the fixed point problem (1.1), where T is a certain operator associated with F. The operator T is usually called the *iteration function*. If we restrict to the real functions of a real single-variable, then one of the most popular algorithms for obtaining T is the well-known *Newton method*, which is based on the iteration function:

$$T(x) = x - \frac{F(x)}{F'(x)}. \tag{1.21}$$

The next theorem gives a recipe for constructing the Newton type method for approximating roots of F.

Theorem 1.5 *Set $F_1 = F$ and, for each $m \geq 2$, recursively, define*

$$F_m(x) := \frac{F_{m-1}(x)}{[F'_{m-1}(x)]^{1/m}}.$$

Then the function

$$G_m(x) = x - \frac{F_{m-1}(x)}{F'_{m-1}(x)}$$

defines an iterative function whose order of the convergence for simple roots is m.

Note that the following results hold:

(1) For $m = 2$, from Theorem 1.5, we obtain the iteration function $G_2 = T$ given in (1.21);
(2) For $m = 3$, we get

$$G_3(x) = x - \frac{F(x)F'(x)}{F'^2(x) - \frac{F(x)F''(x)}{2}},$$

which is the iteration function involved in the Halley method.

(B) The equilibrium problem

In 1966, Hartman and Stampacchia [91] proved the following result:

Theorem 1.6 *Let C be a compact convex subset in $\mathbb{R}^n$ and $f : C \to \mathbb{R}^n$ be a continuous mapping. Then there exists $x_0 \in C$ such that*

$$\langle f(x_0), y - x_0 \rangle \geq 0 \ \ for \ all \ \ y \in C.$$

In 1969, Ky Fan [80] proved the following theorem, which is called *Fan's best approximation theorem*:

Theorem 1.7 *Let C be a nonempty compact convex set in a normed vector space E. Then, for any continuous mapping $f : C \to E$, there exists $x_0 \in C$ such that*

$$\|x_0 - f(x_0)\| = \min_{y \in C} \|y - f(x_0)\|.$$

In particular, if $f(C) \subset C$, then x_0 is a fixed point in f.

Theorems 1.6 and 1.7 play a very important role in nonlinear analysis, variational inequality problems, complementarity problems, optimization problems, game theory, and others.

Let E be a normed vector space with the dual space E^*, C be nonempty convex subset of E, and $f : C \times C \to \mathbb{R}$ be a real-valued bifunction. In 1994, from the ideas of Theorems 1.6 and 1.7, Blum and Oettli [19] considered the *equilibrium problem* (shortly, (EP)) as follows:

Find $x_0 \in C$ such that $f(x_0, y) \geq 0$ for all $y \in C$.

Now, we present the fixed point formula of the equilibrium problem in Hilbert spaces.

Let $\mathscr{H}$ be a real Hilbert space and C be a nonempty closed convex subset of $\mathscr{H}$. We assume the bifunction $f : C \times C \rightarrow \mathbb{R}$ satisfies the following conditions:

(a1) $f(x, x) = 0$ for all $x \in C$;
(a2) f is monotone, that is, $f(x, y) + f(y, x) \leq 0$ for all $x, y \in C$;
(a3) for all $x, y, z \in C$, $\lim_{t \rightarrow 0} f(zt + (1 - t)x, y) \leq f(x, y)$;
(a4) for each fixed $x \in C$, the function $y \longmapsto f(x, y)$ is convex and lower semi-continuous.

Now, $EP(f)$ denotes the set of solutions of the equilibrium problem (EP).

Lemma 1.1 ([19]) *Let $f : C \times C \rightarrow \mathbb{R}$ be a bifunction satisfying the conditions* (a1)–(a4). *Let $r > 0$ and $x \in \mathscr{H}$. Then there exists $z \in C$ such that*

$$f(z, y) + \frac{\langle y - z, z - x \rangle}{r} \geq 0 \ \ for\ all \ y \in C.$$

Lemma 1.2 ([44]) *Let $f : C \times C \rightarrow \mathbb{R}$ be a bifunction satisfying the conditions* (a1)–(a4). *For any $r > 0$ and $x \in \mathscr{H}$, define a mapping $T_r : \mathscr{H} \rightarrow C$ as follows:*

$$T_r(x) = \left\{ z \in C : f(z, y) + \frac{\langle y - z, z - x \rangle}{r} \geq \varepsilon \right\}.$$

Then we have the following:

(1) *T_r is single-valued.*
(2) *T_r is firmly nonexpansive, that is,*

$$\| T_r(x) - T_r(y) \| \leq \langle T_r(x) - T_r(y), x - y \rangle \ \ for\ all \ x, y \in \mathscr{H}.$$

(3) *$Fix(T_r) = EP(f)$.*
(4) *$EP(f)$ is nonempty closed convex.*

The equilibrium problem (EP) has some special cases as follows:

(1) The minimization problem (shortly, (MP))

Find $x \in C$ such that $\phi(x) \leq \phi(y)$ for all $y \in C$,

where $\phi : C \rightarrow \mathbb{R}$ is a function.

If we set $f(x, y) = \phi(y) - \phi(x)$ for all $x, y \in C$, then the problem (EP) is equivalent to the problem (MP).

(2) The saddle point problem (shortly, (SPP))

Find $(\bar{x}_1, \bar{x}_2) \in C_1 \times C_2$ such that $L(\bar{x}_1, y_2) \leq L(\bar{x}_1, \bar{x}_2) \leq L(y_1, \bar{x}_2)$

for all $(y_1, y_2) \in C_1 \times C_2$, where $L : C_1 \times C_2 \to \mathbb{R}$ is a real-valued bifunction. If we set $C = C_1 \times C_2$ and define a bifunction $f : C \times C \to \mathbb{R}$ by

$$f((x_1, x_2), (y_1, y_2)) = L(y_1, x_2) - L(x_1, y_2) \text{ for all } (x_1, x_2), (y_1, y_2) \in C_1 \times C_2,$$

then the problem (EP) coincides with the problem (SPP).

(3) The Nash equilibrium problem (shortly, (NEP))

Let $I = \{1, 2, \cdots, n\}$ be a finite set (a set of players) and, for each $i \in I$, C_i be a strategy set of i-th player. Let $C = \sum_{i \in I} C_i$ and, for each player $i \in I$, $\phi_i : C \to \mathbb{R}$ be a loss function of the i-th player depending on the strategies of all players.

For any $x = (x_1, \cdots, x_n) \in C$, define $x^i = (x_1, \cdots, x_{i-1}, x_{i+1}, \cdots, x_n)$. Then the *Nash equilibrium problem* is formulated as follows:

Find $\bar{x} \in C$ such that, for each $i \in I$, $\phi_i(\bar{x}) \leq \phi_i(\bar{x}^i, y_i)$ for all $y_i \in C_i$,

where $\bar{x} = (\bar{x}_1, \bar{x}_2, \cdots, \bar{x}_n)$ and $(\bar{x}^i, y_i) = (\bar{x}_1, \cdots, \bar{x}_{i-1}, y_i, \bar{x}_{i+1}, \cdots, \bar{x}_n)$. If we define

$$f(x, y) = \sum_{i=1}^{n} (\phi_i(x^i, y_i) - \phi_i(x))$$

then the problem (NEP) is same as the problem (EP).

(4) The fixed point problem (shortly, (FPP))

Let C be a nonempty subset of an inner product space E and $\phi : C \to C$ be a mapping. The *fixed point problem* is formulated as follows

Find $x_0 \in C$ such that $\phi(x_0) = x_0$.

If setting

$$f(x, y) = \langle x - \phi(x), y - x \rangle \text{ for all } x, y \in C,$$

then x_0 is a solution of the problem (FPP) if and only if x_0 is a solution of problem (EP).

(5) The variational inequality problem (shortly, (VIP))

Let E be a normed vector space with the dual space E^* and C be nonempty convex subset of E. Let $\phi : C \to E^*$ be a mapping. Then the *variational inequality problem* is formulated as follows:

Find $x_0 \in C$ such that $\langle \phi(x_0), y - x_0 \rangle \geq 0$ for all $y \in C$.

If setting

$$f(x, y) = \langle \phi(x), y - x \rangle \quad \text{for all } x, y \in C,$$

then the problem (VIP) is equivalent to the problem (EP).

(6) The nonlinear complementarity problem (shortly, (NCP))

Let E be a normed vector space with the dual space E^*, C be a nonempty closed convex cone in E, and

$$C^* = \{x^* \in E^* : \langle x^*, y \rangle \ge 0 \text{ for all } y \in C\}$$

be a polar cone of C. Let $\phi : C \to E^*$ be a mapping. Then the *nonlinear complementarity problem* is formulated as follows:

$$\text{Find } x_0 \in C \text{ such that } \phi(x_0) \in C^* \text{ and } \langle \phi(x_0), x_0 \rangle = 0.$$

Remark 1.1

(a) The problem (VIP) is equivalent to the problem (NCP).
(b) If setting

$$f(x, y) = \langle \phi(x), y - x \rangle \quad \text{for all } x, y \in C,$$

then the problems (VIP), (NCP), and (EP) are equivalent.

Also, we can consider the *dual equilibrium problem* including the *Minty type variational inequality problem* as special cases.

(C) The hierarchical fixed point problem and the equilibrium problem

Let T and V be two nonexpansive mappings from C to C, where C is a nonempty closed and convex subset of a Hilbert space $\mathcal{H}$.

Consider the variational inequality problem (VIP) of finding hierarchically a fixed point of a nonexpansive mapping T with respect to another nonexpansive mapping V, which is called the *hierarchical fixed point problem* (HVIP), that is,

$$\text{Find } x^* \in \text{Fix}(T) \text{ such that } \langle x^* - Vx^*, y - x^* \rangle \ge 0 \quad \text{for all } y \in \text{Fix}(T).$$

Equivalently, $x^* = P_{\text{Fix}(T)} Vx^*$, that is, x^* is a fixed point of the nonexpansive mapping $P_{\text{Fix}(T)} V$, where, for a closed convex subset K of $\mathcal{H}$, P_K is the metric projection of $\mathcal{H}$ on K.

For the problem (HVIP), we have the following:

(1) If $V = I$ in the problem (HVIP), the solution set of the problem (HVIP) is just $\text{Fix}(T)$.

(2) A very particular case of the problem (HVIP) occurs when V is a constant mapping, that is, for any $u \in \mathscr{H}$,

Find $x^* \in \text{Fix}(T)$ such that $\langle x^* - u, x - x^* \rangle \geq 0$ for all $x \in \text{Fix}(T)$

or, equivalently, find the fixed point of the mapping T closest to u, that is,

$$x^* = P_{\text{Fix}(T)}u = \text{argmin}_{x \in \text{Fix}(T)} \frac{1}{2}\|u - x\|^2.$$

(3) If we put $C = \text{Fix}(T)$ and $f(x, y) = \langle (I - V)x, y - x \rangle$, then the problem (HVIP) can be rewritten as follows:

Find $x^* \in C$ such that $f(x^*, y) \geq 0$ for all $y \in C$,

which is the problem (EP).

(D) The integral equation

The integral and integro-differential equations play an important role in the class of operator equations that can be naturally reformulated in terms of the fixed point problem. For any functions f and K, we consider a simple integral equation of the form:

$$y(x) = f(x) + \int_0^1 K(x, s, y(x), y(s))ds \quad \text{for all } x \in [0, 1]. \tag{1.22}$$

The equation of the form (1.22) arises in a variety of contexts. Here, we present three examples.

First, the equation in connection with a problem of radiation transfer is as follows:

$$y(x) = 1 + \int_0^1 \frac{sy(s)y(x)}{s + x}\phi(s)ds,$$

where ϕ is given.

Second, a special but important case of Eq. (1.22) is the Urysohn equation:

$$y(x) = 1 + \int_0^1 K(x, s, y(s))ds.$$

Third, one of the most studied integral equations in fixed point problem is the nonlinear Fredholm integral equation:

$$y(x) = f(x) + \lambda \int_0^1 K(x, s, y(s))ds, \tag{1.23}$$

where $\lambda \in \mathbb{R}$ is a given number. The existence of solutions of Eq. (1.23) has been investigated in the literature (see [160] and the references therein).

Now, we seek a continuous solution of the nonlinear Fredholm integral equation (1.23) by reformulating it as the fixed point problem.

We give the following assumptions:

(A1) $K : [0, 1] \times [0, 1] \times I \to \mathbb{R}$ $(I \subset \mathbb{R})$ is a continuous and bounded mapping on its domain, where $K(x, s, z)$ is called the *kernel* of the integral equation;
(A2) K is L-Lipschitz with respect to the third variable, that is, there exists $L > 0$ such that

$$|K(x, s, z_1) - K(x, s, z_2)| \leq L|z_1 - z_2| \quad \text{for each } x, s \in [0, 1] \text{ and } z_1, z_2 \in I;$$

(A3) $f : [0, 1] \to \mathbb{R}$ is continuous;
(A4) $\lambda \in \mathbb{R}$ is a given number;
(A5) $\varphi : [0, 1] \to I$ is the unknown function, supposed to be continuous.

Let X be the space of all functions $\varphi : [0, 1] \to I$ which satisfy the following conditions:

(B1) φ is continuous;
(B2) $\varphi(x) \in I \subset \mathbb{R}$ for each $x \in [0, 1]$.

Recall that $X = C[0, 1]$ endowed with the (Chebyshev) metric:

$$\mathrm{d}(\varphi_1, \varphi_2) = \max_{x \in [0,1]} |\varphi_1(x) - \varphi_2(x)| \quad \text{for all } \varphi_1, \varphi_2 \in X$$

is a complete metric space.

Now, we define on X the operator T as follows:

$$(T\varphi)(x) = f(x) + \lambda \int_0^1 K(x, s, \varphi(s))ds \quad \text{for all } x \in [0, 1]. \tag{1.24}$$

It is obvious that T maps X into itself (the fact that K and f are continuous implies that $T(\varphi)$ is continuous) and hence $T(X) \subset X$. So, the integral equation (1.23) is equivalent to the fixed point problem:

$$\varphi = T(\varphi),$$

where T is defined by (1.24).

It is easy to show that T is $L|\lambda|$-Lipschitz continuous (see, for example, [18] for the details). By choosing λ such that $|\lambda| < \frac{1}{L}$, it follows that T is in fact a strictly contractive operator and, by the Banach fixed point theorem, T has a unique fixed point, which is the unique solution of the integral equation (1.23) and this solution can be obtained by the Picard iteration.

Next, we consider the following Volterra integral equation of the second kind:

$$y(x) = f(x) + \lambda \int_a^x K(x, s, y(s))ds \quad \text{for all } x \in [a, b], \tag{1.25}$$

where K, f, λ, and y are defined similarly to the previous integral equation.

If K is Lipschitz with respect to the third variable, then (1.25) has a unique solution in the set of continuous functions (see [160] and the references therein). By denoting

$$(T\varphi)(x) = f(x) + \lambda \int_a^x K(x, s, \varphi(s))ds \quad \text{for all } x \in [a, b], \tag{1.26}$$

then (1.25) can be written equivalently into the fixed point form:

$$\varphi = T(\varphi).$$

By choosing suitable parameters, T will be a contractive operator (see, for example, [18] for the details). Therefore, the Picard iteration can be used to approximate the unique solution of the equation.

(E) The ordinary differential equation

We consider the initial value problem for a first order ordinary differential equation (ODE)

$$\begin{cases} y' = f(x, y), \\ y(x_0) = y_0. \end{cases} \tag{1.27}$$

The ODE (1.27) can be written equivalently as the Volterra integral equation:

$$y(x) = y_0 + \int_{x_0}^x f(s, y(s))ds.$$

The two-point boundary value problem:

$$\begin{cases} y'' = f(x, y), \\ y(a) = A, \quad y(b) = B \end{cases} \tag{1.28}$$

can be put into the equivalent integral form:

$$y(x) = \frac{x - a}{b - a} B + \frac{b - x}{b - a} A - \int_a^b G(x, s) f(s, y(s))ds,$$

where $G : [a, b] \times [a, b] \to \mathbb{R}$ given by

$$G(x,s) = \begin{cases} \dfrac{(s-a)(b-x)}{b-a}, & \text{if } a \le s \le x \le b \\[2mm] \dfrac{(x-a)(b-s)}{b-a}, & \text{if } a \le x \le s \le b \end{cases} \qquad (1.29)$$

is the Green function associated with the homogeneous problem:

$$y'' = 0, \quad y(a) = 0, \quad y(b) = 0.$$

Under appropriate assumptions on f (continuous and Lipschitz with respect to the last variable), it is an easy task to show that the corresponding integral operators fulfill a certain contractive condition and hence an appropriate fixed point technique can be used to study these equations under the fixed point formulation.

1.3   Outline

Since its inception in 1953, the Krasnosel'skiǐ–Mann iteration has been one of the highlights of the fixed point theory and a great deal of researchers have studied its convergence in different spaces and presented different variants. Especially, in the last decade, increasing attention has been paid on the Krasnosel'skiǐ–Mann iteration as an important tool in optimization and its accelerations, such as inertia and relaxation, are the focus of the research.

Chapter 3 is devoted to the original Krasnosel'skiǐ–Mann iteration, i.e., the mean iteration scheme introduced by Mann [141], which was recently extended to a broad class of relaxation fixed point algorithms on the convex hull of orbits [46] or the affine hull of orbits [45]. The Krasnosel'skiǐ–Mann iterations with perturbations are given and the bounded perturbation resilience of the Krasnosel'skiǐ–Mann iteration is discussed. This chapter summaries the recent research on the convergence rate of the Krasnosel'skiǐ–Mann iteration, including the global pointwise and ergodic iteration-complexity bounds.

In Chap. 4, we introduce the relations of the Krasnosel'skiǐ–Mann iteration and the operator splitting methods. Several famous operator splitting methods can be transformed into the Krasnosel'skiǐ–Mann iteration, such as the proximal point method, the forward-backward splitting method, the backward-forward splitting method, the Douglas–Rachford splitting method, the Davis–Yin splitting method and a class of primal-dual splitting methods.

In Chap. 5, we discuss the recent progress of the inertial Krasnosel'skiǐ–Mann iteration firstly introduced by Mainge [134], especially, for the general inertial Krasnosel'skiǐ–Mann iteration [58]. We focus on the inertial parameters and present several ranges of inertial parameters. The alternated inertial Krasnosel'skiǐ–Mann iteration and the online inertial Krasnosel'skiǐ–Mann iteration are also introduced.

Chapter 6 considers the multi-step inertial Krasnosel'skiĭ–Mann iteration [65, 67] and presents its convergence analysis. We give two class of inertial parameter sequences which do not involve the iterative sequence. To show the convergence of the corresponding multi-step inertial Krasnosel'skiĭ–Mann iteration, we introduce a general Krasnosel'skiĭ–Mann iteration on the affine hull of orbits and a modified Krasnosel'skiĭ–Mann iteration. As applications, we introduce the multi-step inertial forward-backward splitting method, the multi-step inertial backward-forward splitting method, the multi-step inertial Douglas–Rachford splitting method, the multi-step inertial Davis–Yin splitting method, and the multi-step inertial primal-dual splitting method.

The choice of the relaxation parameters of the Krasnosel'skiĭ–Mann iteration is discussed and we mainly introduce the line search method, the online choice, and the optimal choice of the relaxation parameters in Chap. 7. Based on the equivalent relation of the fixed point problems and variational inequalities problems, we present some (multi-step) inertial Krasnosel'skiĭ–Mann iterations where the relaxation parameters are given by adaptive ways. We also introduce a residual algorithm, by using the equivalence of the fixed point problems and nonlinear monotone equations, which is in fact the Krasnosel'skiĭ–Mann iteration with adaptive relaxation parameters.

The final chapter is devoted to two applications of the Krasnosel'skiĭ–Mann iteration, including asynchronous parallel coordinate updates methods and cyclic coordinate-update algorithms. Their convergence analysis and convergence rate are presented under some conditions.

Chapter 2
Notation and Mathematical Foundations

In this chapter, we present some concepts, definitions, and lemmas which will be used in the following chapters.

(A) Notation

Throughout the paper, $\mathbb{N}$ denotes the set of nonnegative integers, and $\mathscr{H}$ and $\mathscr{G}$ denote real Hilbert spaces with the scalar product $\langle \cdot, \cdot \rangle$, and the norm $\| \cdot \|$. $\mathbb{R}^N$ denotes the standard N-dimensional Euclidean space. Forward, we use the notation:

- $\rightharpoonup$ for the weak convergence and $\to$ for the strong convergence;
- $\omega_w(x_n) = \{x : \exists x_{n_j} \rightharpoonup x\}$ denotes the weak ω-limit set of the sequence $\{x_n\}_{n \in \mathbb{N}}$;
- ℓ_+^1 denotes the set of summable sequences in $[0, +\infty)$;
- The positive and negative parts of $\xi \in \mathbb{R}$ are denoted by $\xi^+ = \max\{0, \xi\}$ and $\xi^- = -\min\{0, \xi\}$, respectively.

(B) Nonlinear operators

We recall the definitions of some nonlinear operators and discuss their properties and relations.

Definition 2.1 An operator $T : \mathscr{H} \to \mathscr{H}$ is said to be:

(1) *L-Lipschitz continuous* for $L \geq 0$ if

$$\|T(x) - T(y)\| \leq L\|x - y\| \quad \text{for all } x, y \in \mathscr{H}.$$

(2) *nonexpansive* if $L = 1$.
(3) *contractive* if $0 < L < 1$.

Definition 2.2 An operator $T : \mathscr{H} \to \mathscr{H}$ is said to be *pseudocontractive* if

$$\|T(x) - T(y)\|^2 \leq \|x - y\|^2 + \|(\text{Id} - T)x - (\text{Id} - T)y\|^2 \quad \text{for all } x, y \in \mathscr{H}.$$

Q.-L. Dong et al., *The Krasnosel'skiĭ-Mann Iterative Method*, SpringerBriefs in Optimization, https://doi.org/10.1007/978-3-030-91654-1_2

It is easy to see that the class of pseudocontractive operators includes the class of nonexpansive operators.

Definition 2.3 An operator $T : \mathscr{H} \to \mathscr{H}$ with a fixed point is said to be:

(1) *k-demicontractive* for $k \in \mathbb{R}$ if

$$\|T(x) - y\|^2 \leq \|x - y\|^2 + k\|T(x) - x\|^2 \quad \text{for all } x \in \mathscr{H} \text{ and } y \in \text{Fix}(T).$$

(2) *quasi-nonexpansive* if $k = 0$.

It is obvious that if the operator T with a fixed point is nonexpansive, then it is quasi-nonexpansive, but the reverse is not necessarily true.

Definition 2.4 An operator $T : \mathscr{H} \to \mathscr{H}$ is said to be *firmly nonexpansive* if

$$\|T(x) - T(y)\|^2 \leq \langle T(x) - T(y), x - y \rangle \quad \text{for all } x, y \in \mathscr{H}.$$

If T is firmly nonexpansive, then it is nonexpansive.

Lemma 2.1 ([15, Proposition 4.2] and [151]) *Let $T : \mathscr{H} \to \mathscr{H}$ be an operator. Then the following statements are equivalent:*

(1) *T is firmly nonexpansive.*
(2) *$2T - \text{Id}$ is nonexpansive.*

Averaged operators introduced in [12] is one of the most important class of non-linear operators and their central role in many nonlinear analysis and optimization algorithms was pointed out in [42] with further refinements in [196].

Definition 2.5 Let $T : \mathscr{H} \to \mathscr{H}$ be an operator. Then, for any $\alpha \in (0, 1)$, T is said to be *α-averaged* if there exists a nonexpansive operator $S : \mathscr{H} \to \mathscr{H}$ such that

$$T = \alpha S + (1 - \alpha)\text{Id}.$$

Remark 2.1 If T is firmly nonexpansive, then T is $\frac{1}{2}$-averaged.

Lemma 2.2 ([38, Theorem 3(b)] and [50, Proposition 2.4]) *Let C be a nonempty subset of $\mathscr{H}$, $T_1 : C \to C$ be σ_1-averaged, and $T_2 : C \to C$ be σ_2-averaged for any $\sigma_1, \sigma_2 \in (0, 1)$. Set $T = T_1 T_2$ and*

$$\sigma = \frac{\sigma_1 + \sigma_2 - 2\sigma_1\sigma_2}{1 - \sigma_1\sigma_2}. \tag{2.1}$$

Then $\sigma \in (0, 1)$ and T is σ-averaged.

Remark 2.2 Besides (2.1), there are other two averaged constants for the composition of two averaged operators, i.e.,

$$\sigma = \frac{2\max\{\sigma_1, \sigma_2\}}{\max\{\sigma_1, \sigma_2\} + 1} \quad \text{and} \quad \sigma = \sigma_1 + \sigma_2 - \sigma_1\sigma_2.$$

The smallest (best) averagedness constant is key to the relaxation parameters of the corresponding method (see, for example, [114, 197]). Combettes and Yamada [50] compared these averaged constants and showed that the averaged constant given by (2.1) is the smallest. Very recently, Huang et al. [108] proved that the averaged constant given by (2.1) is tight.

Definition 2.6 Let $S : \mathscr{H} \to \mathscr{H}$ and let $\alpha \in (0, +\infty)$. Then the operator $T = \text{Id} + \alpha(S - \text{Id})$ is called a *relaxation* of S. Furthermore, we have the following:

(1) If $\alpha \leq 1$, then T is an *underrelaxation* of S.
(2) If $\alpha > 1$, then T is an *overrelaxation* of S.
(3) In particular, if $\alpha = 2$, then T is the *reflection* of S.

From Definitions 2.5 and 2.6, it follows that an averaged operator is an underrelaxation of one nonexpansive operator.

Definition 2.7 Let X be a Banach space with the dual space X^* and $J : X \to X^*$ be the normalized duality map. Let C be a nonempty closed convex subset of X. Let $\varphi : [0, \infty) \to [0, \infty)$ be a continuous, strictly increasing function with $\varphi(0) = 0$ and $\lim_{t \to \infty} \varphi(t) = \infty$. An operator $T : C \to X$ is said to be:

(1) *accretive* (see [57]) if, for each $x, y \in C$,

$$\langle T(x) - T(y), J(x - y) \rangle \geq 0.$$

(2) *φ-accretive* if, for each $x, y \in C$,

$$\langle T(x) - T(y), J(x - y) \rangle \geq [\varphi(\|x\|) - \varphi(\|y\|)](\|x\| - \|y\|).$$

Definition 2.8 ([170]) A mapping $T : \mathscr{H} \to \mathscr{H}$ is said to be *demicompact* at $x \in \mathscr{H}$ if, for every bounded sequence $\{x_n\}$ in $\mathscr{H}$ such that $T(x_n) - x_n \to 0$, there exists a strongly convergent subsequence.

Definition 2.9 Let C be a nonempty closed and convex subset of $\mathscr{H}$. For each point $x \in \mathscr{H}$, there exists a unique nearest point in C, denoted by $P_C(x)$, that is,

$$\|x - P_C(x)\| \leq \|x - y\| \quad \text{for all } y \in C. \tag{2.2}$$

The mapping $P_C : \mathscr{H} \to C$ is called the *metric projection* of $\mathscr{H}$ onto C.

The characterization of the metric projection P_C is given in the next lemma:

Lemma 2.3 ([88, Section 3]) *Let $x \in \mathscr{H}$ and $z \in C$. Then $z = P_C(x)$ if and only if*

$$\langle x - z, z - y \rangle \geq 0 \ \text{for all } y \in C. \tag{2.3}$$

It is well known that P_C is a firmly nonexpansive mapping of $\mathcal{H}$ onto C. This is captured in the next lemma:

Lemma 2.4 *For any* $x, y \in \mathcal{H}$ *and* $z \in C$, *the following hold:*

(1) $\|P_C(x) - P_C(y)\|^2 \leq \langle P_C(x) - P_C(y), x - y \rangle$.
(2) $\|P_C(x) - z\|^2 \leq \|x - z\|^2 - \|P_C(x) - x\|^2$.
(3) $\langle (\mathrm{Id} - P_C)x - (\mathrm{Id} - P_C)y, x - y \rangle \geq \|(\mathrm{Id} - P_C)x - (\mathrm{Id} - P_C)y\|^2$.

Lemma 2.5 *Let* $T : \mathcal{H} \to \mathcal{H}$ *be nonexpansive. Then* $\mathrm{Fix}(T)$ *is closed and convex.*

The following result, which is a consequence of the demiclosedness principle, is useful in the proof of the convergence of the Krasnosel'skiĭ–Mann iteration:

Lemma 2.6 ([15, Corollary 4.18]) *Let* $C \subseteq \mathcal{H}$ *be nonempty closed and convex and* $T : C \to \mathcal{H}$ *be a nonexpansive mapping. Let* $\{x_n\}$ *be a sequence in* C *and let* $x \in \mathcal{H}$ *such that* $x_n \rightharpoonup x$ *and* $T(x_n) - x_n \to 0$ *as* $n \to +\infty$. *Then* $x \in \mathrm{Fix}(T)$.

The next lemma is key in proving the weak convergence of the Krasnosel'skiĭ–Mann iteration:

Lemma 2.7 ([15, Lemma 2.39]) *Let* C *be a nonempty subset of* $\mathcal{H}$ *and* $\{x_n\}$ *be a sequence in* $\mathcal{H}$ *such that the following two conditions hold:*

(a) *for all* $x \in C$, $\lim_{n\to\infty} \|x_n - x\|$ *exists;*
(b) *every sequential weak cluster point of the sequence* $\{x_n\}$ *is in* C.

Then the sequence $\{x_n\}$ *converges weakly to a point in* C.

Lemma 2.8 ([60, Lemma 2.6]) *Assume that* $\{a_n\}$ *is a sequence of nonnegative real numbers such that*

$$a_{n+1} \leq (1 + \gamma_n)a_n + \delta_n \quad \text{for each } n \geq 0,$$

where the sequences $\{\gamma_n\} \subset [0, +\infty)$ *and* $\{\delta_n\}$ *satisfy the following conditions:*

(a) $\sum_{n=0}^{\infty} \gamma_n < +\infty$;
(b) $\sum_{n=0}^{\infty} \delta_n < +\infty$ *or* $\limsup_{n\to\infty} \delta_n \leq 0$.

Then $\lim_{n\to\infty} a_n$ *exists.*

Using the Cauchy–Schwarz inequality and the mean value inequality, we easily show the following lemma:

Lemma 2.9 *For any* $a, b \in \mathcal{H}$, *it holds:*

$$\|a - b\|^2 \leq (1 + \|b\|)\|a\|^2 + \|b\| + \|b\|^2.$$

The following identity is used several times in the paper (see Corollary 2.15 of [15]): for all $\alpha \in \mathbb{R}$ and $(x, y) \in \mathcal{H} \times \mathcal{H}$,

$$\|\alpha x + (1 - \alpha)y\|^2 = \alpha \|x\|^2 + (1 - \alpha)\|y\|^2 - \alpha(1 - \alpha)\|x - y\|^2. \tag{2.4}$$

(C) Convex analysis

Let $A : \mathscr{H} \rightrightarrows \mathscr{H}$ be a set-valued operator. Then we have the following notations:

(1) The *domain* of A is $\operatorname{dom} A = \{x \in \mathscr{H} : Ax \neq \emptyset\}$.
(2) The *range* of A is $\operatorname{ran} A = \{y \in \mathscr{H} : \exists x \in \mathscr{H}, y \in Ax\}$.
(3) The *graph* of A is the set $\operatorname{gra} A = \{(x, y) \in \mathscr{H}^2 : y \in Ax\}$.
(4) The *inverse* of A is the operator whose graph is

$$\operatorname{gra} A^{-1} = \{(y, x) \in \mathscr{H}^2 : x \in A^{-1}y\}.$$

(5) The set of *zeros* is $\operatorname{zer} A = \{x \in \mathscr{H} : 0 \in Ax\} = A^{-1}(0)$.

Definition 2.10

(1) A set-valued operator $A : \mathscr{H} \rightrightarrows \mathscr{H}$ is said to be *monotone* if

$$\langle x - y, u - v \rangle \geq 0 \quad \text{for all } (x, u), (y, v) \in \operatorname{gra} A.$$

(2) The operator $A : \mathscr{H} \rightrightarrows \mathscr{H}$ is said to be *maximally monotone* if $\operatorname{gra} A$ is not strictly contained in the graph of any other monotone operator.

Definition 2.11 Let $A : \mathscr{H} \rightrightarrows \mathscr{H}$ be a maximally monotone operator.

(1) For any $\gamma > 0$, the *resolvent* of A is defined by

$$J_{\gamma A} = (\operatorname{Id} + \gamma A)^{-1}. \tag{2.5}$$

(2) For any $\gamma > 0$, the *reflection resolvent* (see [7, 151]) of A is defined by

$$R_{\gamma A} = 2 J_{\gamma A} - \operatorname{Id}. \tag{2.6}$$

(3) For $\gamma \in \mathbb{R}$, the *Yosida approximation* of A with γ is defined by

$$A_\gamma = (A^{-1} + \gamma \operatorname{Id})^{-1}.$$

Lemma 2.10 ([15, Proposition 23.7 and Corollary 23.10]) *Let* $A : \mathscr{H} \rightrightarrows \mathscr{H}$ *be a maximally monotone operator and* $T : \mathscr{H} \to \mathscr{H}$ *be a firmly nonexpansive mapping. Then we have the following:*

(1) *J_A is firmly nonexpansive.*
(2) *there exists a maximally monotone operator* $A' : \mathscr{H} \rightrightarrows \mathscr{H}$ *such that* $T = J_{A'}$.

By Lemmas 2.1(2) and 2.10(1), the reflection resolvent $R_{\gamma A}$ is nonexpansive. However, it is not firmly nonexpansive or averaged.

Definition 2.12 An operator $B : \mathscr{H} \to \mathscr{H}$ is said to be *β-cocoercive* for some $\beta \in (0, +\infty)$ if

$$\beta \|Bx - By\|^2 \le \langle Bx - By, x - y \rangle \quad \text{for all} \ x, y \in \mathscr{H}.$$

Observe that the β-cocoercivity implies the $\frac{1}{\beta}$-Lipschitz continuity. If T is nonexpansive, then $\mathrm{Id} - T$ is $\frac{1}{2}$-cocoercive. The converse is also true.

Lemma 2.11 *Let operator* $B : \mathscr{H} \to \mathscr{H}$ *be* β*-cocoercive for some* $\beta > 0$. *Then we have the following:*

(1) βB *is firmly nonexpansive.*
(2) $\mathrm{Id} - \gamma B$ *is* $\frac{\gamma}{2\beta}$*-averaged for* $\gamma \in (0, 2\beta)$.

Definition 2.13 An operator $A : \mathscr{H} \rightrightarrows \mathscr{H}$ is said to be *uniformly monotone* if there exists an increasing function $\phi_A : [0, +\infty) \to [0, +\infty)$ that vanishes only at 0 and satisfies

$$\langle x - y, u - v \rangle \ge \phi_A(\|x - y\|) \quad \text{for all} \ (x, u), (y, v) \in gra A.$$

A prominent representative of the class of uniformly monotone operators is the strongly monotone operator.

Definition 2.14 Let $\gamma > 0$ be arbitrary.

(1) A mapping $A : \mathscr{H} \to \mathscr{H}$ is said to be γ*-strongly monotone* if

$$\langle x - y, Ax - Ay \rangle \ge \gamma \|x - y\|^2 \quad \text{for all} \ x, y \in \mathscr{H}.$$

(2) A mapping $A : \mathscr{H} \to \mathscr{H}$ is said to be *quasi-*γ*-strongly monotone* if

$$\langle x - y, Ax \rangle \ge \gamma \|x - y\|^2 \quad \text{for all} \ x \in \mathscr{H} \ \text{and} \ y \in \mathrm{zer}(A).$$

If the operator A is γ-strongly monotone with $\mathrm{zer}(A) \ne \emptyset$, then it is quasi-$\gamma$-strongly monotone.

Definition 2.15 (Parallel Sum) Let $C, D : \mathscr{H} \rightrightarrows \mathscr{H}$ be two set-valued operators. The *parallel sum* of C and D is defined by

$$C \square D := (C^{-1} + D^{-1})^{-1}.$$

Definition 2.16

(1) A function $f : \mathscr{H} \to (-\infty, +\infty]$ is said to be *convex* if, for all $x, y \in \mathscr{H}$ and $\alpha \in (0, 1)$,

$$f(\alpha x + (1 - \alpha)y) \le \alpha f(x) + (1 - \alpha) f(y).$$

(2) A function $f : \mathscr{H} \to (-\infty, +\infty]$ is said to be *lower semi-continuous* if, for every sequence $\{x_n\}_{n \in \mathbb{N}}$ in $\mathscr{H}$ and $x \in \mathscr{H}$,

$$x_n \rightarrow x \implies f(x) \leq \underline{\lim}_{n \to \infty} f(x_n).$$

(3) A function $f : \mathcal{H} \rightarrow (-\infty, +\infty]$ is said to be *proper* if

$$\text{dom} \, f = \{x \in \mathcal{H} : f(x) < +\infty\} \neq \emptyset.$$

Denote by $\Gamma_0(\mathcal{H})$ the set of proper lower semi-continuous convex functions from $\mathcal{H}$ to $(-\infty, +\infty]$.

Definition 2.17 Let the function $f \in \Gamma_0(\mathcal{H})$.

(1) The *subdifferential* of f is a set-valued operator defined as follows: for all $x \in \mathcal{H}$,

$$\partial f(x) := \{g \in \mathcal{H} : \langle x' - x, g \rangle + f(x) \leq f(x'), \, \forall x' \in \mathcal{H}\}.$$

(2) A function f is said to be *subdifferentiable* at x if $\partial f(x) \neq \emptyset$.
(3) An element of $\partial f(x)$ is called a *subgradient*.

The subdifferential is a generalization of the derivative to non-differentiable functions and it is always a convex closed set. The relation of the subdifferential and the maximally monotone operator is given as follows:

Lemma 2.12 *If $f \in \Gamma_0(\mathcal{H})$, then ∂f is maximally monotone.*

Let C be a nonempty closed convex subset of $\mathcal{H}$. Then the *indicator function* ι_C is defined by

$$\iota_C(x) = \begin{cases} 0, & \text{if } x \in C, \\ +\infty, & \text{otherwise.} \end{cases}$$

The *normal cone operator* N_C of C is the subdifferential of ι_C, i.e., $N_C = \partial \iota_C$ is defined by

$$N_C(x) = \begin{cases} \{u \in \mathcal{H} : \langle y - x, u \rangle \leq 0, \, \forall y \in C\}, & \text{if } x \in C, \\ \emptyset, & \text{otherwise.} \end{cases}$$

Given a function, Fermat's theorem adequately characterizes its extrema. For any convex function, we have the following result:

Theorem 2.1 (Fermat's Theorem) *If $f : \mathcal{H} \rightarrow (-\infty, +\infty]$ is a proper convex mapping, then x_*^* is a minimizer of f if and only if*

$$0 \in \partial f(x^*) \tag{2.7}$$

The condition (2.7) is also called the *first-order (necessary and sufficient) optimality condition*.

Definition 2.18 Let $f \in \Gamma_0(\mathcal{H})$ and $\gamma > 0$ be a parameter. For any $x \in \mathcal{H}$, the *proximity operator* (or *proximal mapping*) of f parameterized by γ is defined by

$$\text{prox}_{\gamma f}(x) := \underset{y \in \mathcal{H}}{\text{argmin}}\, f(y) + \frac{1}{2\gamma}\|y - x\|^2.$$

Note that $\text{prox}_{\gamma f}$ is firmly nonexpansive. A very simple and widely used example of the proximity operator is the projection operator. Let C be a nonempty closed convex set of $\mathcal{H}$, we have $\text{prox}_{\iota_C} = P_C$.

Let $f \in \Gamma_0(\mathcal{H})$ and ∂f be the subdifferential of f. Combining together (2.5), Definition 2.18 and Lemma 2.12, we have

$$\text{prox}_{\gamma f} = J_{\gamma \partial f},$$

where $\gamma > 0$ (see also [152]). Then, in terms of the set of fixed points, we have

$$\text{Fix}(\text{prox}_{\gamma f}) = \text{Fix}(J_{\gamma \partial f}) = \text{zer}(\partial f).$$

Chapter 3
The Krasnosel'skiĭ–Mann Iteration

In this chapter, we discuss the progress of the original Krasnosel'skiĭ–Mann iteration, the perturbations of the Krasnosel'skiĭ–Mann iteration, and several convergence rates.

3.1 The Original Krasnosel'skiĭ–Mann Iteration

Let X be a normed space and C be a convex subset of X. Let $T : C \to C$ be an operator. Given a real sequence $\{\lambda_n\} \subset [0, 1]$, the *iterative scheme* is addressed as

$$x_{n+1} = (1 - \lambda_n)x_n + \lambda_n T(x_n) \quad \text{for each } n \geq 0, \tag{3.1}$$

starting by $x_0 \in C$.

The above iteration has been examined in a very large amount of papers in the last sixty years. It is well known as the *Krasnosel'skiĭ–Mann iteration* since Mann in 1953 firstly investigated it and Krasnosel'skiĭ in 1955 examined the particular case $\lambda_n = \lambda$ for each $n \geq 0$. Some other authors called it the *Mann–Toeplitz process* since Toeplitz studied a more general matrix approach for which (3.1) is a particular case.

Next, we recall the iterative scheme in the original paper of Mann [141]. First, we introduce an infinite lower triangular matrix A, i.e.,

$$A = \begin{bmatrix} 1 & 0 & 0 & \cdots & 0 & \cdots \\ a_{21} & a_{22} & 0 & \cdots & 0 & \cdots \\ \cdots & \cdots & \cdots & \cdots & \cdots & \cdots \\ a_{n1} & a_{n2} & \cdots & \cdots & a_{nn} & \cdots \\ \cdots & \cdots & \cdots & \cdots & \cdots & \cdots \end{bmatrix},$$

Q.-L. Dong et al., *The Krasnosel'skiĭ-Mann Iterative Method*, SpringerBriefs in Optimization, https://doi.org/10.1007/978-3-030-91654-1_3

whose elements satisfy the following restrictions:

(A1) $a_{nk} \geq 0$ for each n, $k \geq 1$ and $a_{nk} = 0$ for each $k > n$;
(A2) $\sum_{k=1}^{n} a_{nk} = 1$ for each $n \geq 1$.

Starting with an arbitrary element $x_1 \in C$, Mann introduced the following iteration process:

$$\begin{cases} v_n = \displaystyle\sum_{k=1}^{n} a_{nk} x_k, \\[2mm] x_{n+1} = T(v_n). \end{cases} \tag{3.2}$$

It is observed that the $(n+1)$st element in (3.2) is the image under T of the centroid of the first n elements.

This process is completely determined by the initial point x_1, the matrix A, and the mapping T. For this reason, it can be denoted briefly by (x_1, A, T) and regarded as a generalized iteration process because when A is the identity matrix I the process (x_1, I, T) is just the Picard iteration (1.2).

Let C be a convex compact set in a Banach space and $T : C \to C$ be a nonexpansive mapping. Then Mann [141] showed the following result:

Theorem 3.1 *If either of the sequences $\{x_n\}$ and $\{v_n\}$ converges, then the other also converges to the same point and their common limit is a fixed point of T.*

As a particular case of the general process (x_1, A, T), Mann considered the Cesàro matrix:

$$A = \begin{bmatrix} 1 & 0 & 0 & \cdots 0 & \cdots \\ \frac{1}{2} & \frac{1}{2} & 0 & \cdots 0 & \cdots \\ \frac{1}{3} & \frac{1}{3} & \frac{1}{3} & \cdots 0 & \cdots \\ \cdots & \cdots & \cdots & \cdots & \cdots \\ \frac{1}{n} & \frac{1}{n} & \cdots & \cdots \frac{1}{n} & \cdots \\ \cdots & \cdots & \cdots & \cdots & \cdots \end{bmatrix}.$$

The process (3.2) corresponding to A is

$$\begin{cases} v_n = \dfrac{1}{n} \displaystyle\sum_{k=1}^{n} x_k, \\[2mm] x_{n+1} = T(v_n). \end{cases} \tag{3.3}$$

It is easy to show that in this process

$$v_{n+1} = \left(1 - \frac{1}{n+1}\right) v_n + \frac{1}{n+1} T(v_n). \tag{3.4}$$

Mann obtained the convergence result for the continuous function defined in the real axis:

Theorem 3.2 ([141, Theorem 4]) *If $T : [a, b] \to [a, b]$ is a continuous function having a unique fixed point p, then the sequence $\{x_n\}$ defined by the iteration (3.3) converges to p for all choices of $x_1 \in [a, b]$.*

Thereafter, Franks and Marzec [82] removed the uniqueness of fixed points of T in Theorem 3.2. Borwein and Borwein [20] proved the further results. Let $\{d_n\}$ be a nonnegative sequence satisfying $d_1 > 0$ and $\sum_{k=1}^{+\infty} d_k = +\infty$ and let

$$a_{nk} = \frac{d_k}{\sum_{k=1}^{n} d_k}.$$

Theorem 3.3 ([20, Theorem 11]) *If $T : [a, b] \to [a, b]$ is a function with $Fix(T) \neq \emptyset$. Then the sequence $\{x_n\}$ defined by the iteration (3.2) converges to a fixed point of T provided that one of the following conditions is satisfied:*

(a) *T is continuous and*

$$\lim_{n \to \infty} a_{nn} = 0;$$

(b) *T is L-Lipschitz continuous and*

$$\limsup_{n \to \infty} a_{nn} \leq \frac{2}{L + 1}.$$

It is easy to see that the Cesàro matrix satisfies the condition (a) of Theorem 3.3. Note that Theorem 3.3 fails in higher dimensional spaces (see a counterexample in [20]).

Thirteen years later, Dotson [75], probably inspired by the example of Cesàro matrix, defined the normal Mann process as a Mann process (x_1, A, T) for which the matrix A satisfies not only (A1) and (A2) but also

(A3) $\lim_{n \to \infty} a_{nk} = 0$ for each $k \geq 1$;
(A4) $a_{n+1,k} = (1 - a_{n+1,n+1})a_{nk}$ for each $n \geq 1$ and $k = 1, 2, \cdots, n$.

It is verified easily that Cesàro matrix satisfies the properties (A1)–(A4). The property (A4) implies that the sequences $\{v_n\}$ generated by (3.2) can be written as:

$$v_{n+1} = a_{n+1,n+1}x_{n+1} + \sum_{k=1}^{n} a_{n+1,k}x_k$$

$$= a_{n+1,n+1}x_{n+1} + (1 - a_{n+1,n+1}) \sum_{k=1}^{n} a_{n,k}x_k$$

$$= a_{n+1,n+1}Tv_n + (1 - a_{n+1,n+1})v_n.$$

Therefore, one is really just applying (3.1) with a specific relaxation strategy, that is,

$$\lambda_n = a_{n+1,n+1} \text{ for each } n \geq 1,$$

which are the elements of the main diagonal of the matrix A.

For this reason, the Krasnosel'skiĭ–Mann iteration (3.1) is also referred to as the "Mann–Dotson iterates" in the literature (see, for example, [142]) although it merely corresponds to a special case of (3.2).

Combettes and Pennanen [46] considered the inexact Krasnosel'skiĭ–Mann-like generalization for a finite family of operators. The corresponding exact case with one mapping is given as follows: for each $n \geq 0$,

$$\begin{cases} v_n = \displaystyle\sum_{k=0}^{n} a_{nk} x_k, \\ x_{n+1} = (1 - \lambda_n) v_n + \lambda_n T(v_n). \end{cases} \tag{3.5}$$

The iterative scheme (3.5) can be seen as the relaxation version of the mean iteration scheme (3.2) and reduces to the latter when $\lambda_n \equiv 1$.

To guarantee the convergence of the iterative sequence $\{x_n\}$ generated by (3.5), the following definition was introduced:

Definition 3.1 The array $\{a_{nk}\}_{n\in\mathbb{N}, 0 \leq k \leq n}$ is said to be *concentrating* if every sequence $\{\xi_n\}_{n\in\mathbb{N}}$ in $\mathbb{R}_+$ converges under the following condition:

$$\xi_{n+1} \leq \sum_{k=0}^{n} a_{nk} \xi_k \text{ for each } n \geq 0.$$

For convenience, the array $\{a_{nk}\}_{n\in\mathbb{N}, 0 \leq k \leq n}$ being concentrating is called to satisfy the *condition* (A5).

Some arrays satisfying (A1)–(A3) and (A5) were presented in [46], for example, the simplest of which is $\{a_{nk}\}_{n\in\mathbb{N}, 0 \leq k \leq n}$ with $a_{nn} = 1$ for each $n \in \mathbb{N}$ and $a_{nk} = 0$ for $0 \leq k < n$ for each $n \in \mathbb{N}$. In this case, (3.5) reverts to (3.1). The following are two simple examples:

Example 3.1 Let $a_{nn} = a$ for each $n \geq 0$, where $a \in (0, 1)$. Set

$$a_{nk} = \begin{cases} (1 - a)^n, & \text{if } k = 0, \\ a(1 - a)^{n-k}, & \text{if } 1 \leq k \leq n. \end{cases}$$

Example 3.2 Let $\{\alpha_l\}_{0 \leq l \leq m}$ be a sequence of strictly positive numbers such that $\sum_{l=0}^{m} \alpha_l = 1$. Set,
for each $n \in \{0, \cdots, m - 1\}$,

$$a_{nk} = \begin{cases} 0, & \text{if } 0 \le k < n, \\ 1, & \text{if } k = n; \end{cases} \tag{3.6}$$

for each $n \ge m$,

$$a_{nk} = \begin{cases} 0, & \text{if } 0 \le k < n - m, \\ \alpha_{n-k}, & \text{if } n - m \le k \le n. \end{cases} \tag{3.7}$$

The convergence result of the scheme (3.5) for the firmly nonexpansive mapping T is given as follows:

Theorem 3.4 ([46, Theorem 3.5]) *Let $\{\lambda_n\}$ lie in $[\delta, 2 - \delta]$ for some $\delta \in (0, 2)$. Suppose that the array $\{a_{nk}\}_{n \in \mathbb{N}, 0 \le k \le n}$ satisfies (A1)–(A3) and (A5) and $T : \mathscr{H} \to \mathscr{H}$ is a firmly nonexpansive mapping with $\mathrm{Fix}(T) \ne \emptyset$. Then the sequence $\{x_n\}$ generated by the scheme (3.5) weakly converges to a fixed point of T.*

In fact, the scheme (3.5) can be seen as an iteration on the convex hull of orbits since $\{a_{nk}\}_{n \in \mathbb{N}, 0 \le k \le n}$ satisfies (A1) and (A2) which means that v_n is a convex combination of the previous iterates $x_0, x_1, \cdots, x_n$.

Recently, Combettes and Glaudin [45] got rid of the condition (A1) and considered a more general case: the iteration on the affine hull of orbits. They made the assumptions for the array $\{a_{nk}\}_{n \in \mathbb{N}, 0 \le k \le n}$ as follows:

(B1) $\sup_{n \ge 0} \sum_{k=0}^{n} |a_{nk}| < +\infty$;
(B2) $\sum_{k=0}^{n} a_{nk} = 1$ for all $n \ge 0$;
(B3) for all $k \ge 0$, $\lim_{n \to +\infty} a_{nk} = 0$;
(B4) there exists a sequence $\{\chi_n\}$ in $(0, +\infty)$ such that $\inf_{n \ge 0} \chi_n > 0$ and every sequence $\{\xi_n\}$ in $[0, +\infty)$ converges under the following condition:

$$\left(\exists \{\varepsilon_k\} \in [0, +\infty)^{\mathbb{N}} \right) \begin{cases} \displaystyle\sum_{n=0}^{+\infty} \chi_n \varepsilon_n < +\infty, \\ (\forall k \in \mathbb{N}) \ \ \xi_{n+1} \le \displaystyle\sum_{k=0}^{n} a_{nk} \xi_k. \end{cases} \tag{3.8}$$

For the nonnegative real array, the conditions (B1)–(B4) reduce the conditions (A1)–(A3) and (A5).

Combettes and Glaudin [45] constructed an instance of an array $\{a_{nk}\}_{n \in \mathbb{N}, 0 \le k \le n}$ satisfying the conditions (B1)–(B4) with negative entries.

Example 3.3 ([45, Example 2.5]) Let $\{a_{nk}\}_{n \in \mathbb{N}, 0 \le k \le n}$ be a real array such that $a_{11} = 1$ and, for each $n \ge 0$,

$$1 \le a_{nn} < 2 \quad \text{and} \quad a_{nk} = \begin{cases} 1 - a_{nn}, & \text{if } k = n - 1, \\ 0, & \text{if } 1 \le k < n - 1. \end{cases}$$

From [45, Theorem 3.1], it is easy to obtain the following convergence result:

Theorem 3.5 *Let* $T : \mathscr{H} \to \mathscr{H}$ *be a nonexpansive mapping with* $\mathrm{Fix}(T) \ne \emptyset$. *Assume that* $\{a_{nk}\}_{n \in \mathbb{N}, 0 \le k \le n}$ *satisfies Assumptions* (B1)–(B4) *and* $\{\lambda_n\}$ *lies in* $[\delta, 1 - \delta]$ *for some* $\delta \in (0, 1)$. *Let* $\{x_n\}$ *be the sequence generated by* (3.5) *which is assumed to satisfy*

$$\sum_{n=0}^{+\infty} \chi_n \sum_{k=0}^{n} \sum_{l=0}^{k} [a_{nk} a_{nl}]^- \|x_k - x_l\|^2 < +\infty. \tag{3.9}$$

Then the sequence $\{x_n\}$ *converges weakly to a point in* $\mathrm{Fix}(T)$.

As far as we know, except for the matrices satisfying the conditions of Theorem 3.3 there is no matrix $\{a_{nk}\}_{n \ge 1, k \ge 1}$ for which the mean iteration scheme (3.2) converges. To this end, we propose the following problem:

Open Problem *Except for the matrices satisfying the conditions of Theorem 3.3 whether does there exist general matrix* $\{a_{nk}\}_{n \ge 1, k \ge 1}$ *for which the mean iteration scheme* (3.2) *converges?*

3.2 Some Results on the Weak and Strong Convergence

Let X be a Banach space, C be a closed convex subset of X, and $T : C \to C$ be a nonexpansive mapping with $\mathrm{Fix}(T) \ne \emptyset$.

Let $\{x_n\}$ be the sequence generated by the Krasnosel'skiĭ–Mann iteration (3.1). Then the following hold:

(i) $\|x_{n+1} - p\| \le \|x_n - p\|$ for each $n \ge 0$ and $p \in \mathrm{Fix}(T)$,
(ii) $\|x_{n+1} - T(x_{n+1})\| \le \|x_n - T(x_n)\|$ for each $n \ge 0$.

Note that (i) and (ii) are easy to verify, which are the basic properties of the sequence generated by Krasnosel'skiĭ–Mann iteration (3.1). From (ii), it follows that $\lim_{n \to \infty} \|x_n - T(x_n)\|$ exists. However, it is not sufficient to obtain the convergence of the sequence $\{x_n\}$ by using (i) and (ii). We further need the condition

$$\lim_{n \to \infty} \|x_n - T(x_n)\| = 0.$$

Borwein et al. [21] also presented the following properties:

(iii) $\|x_{n+1} - x_{n+1}^*\| \le \|x_n - x_n^*\|$, where $\{x_n^*\}$ is also a sequence generated by Krasnosel'skiĭ–Mann iteration with initial starting point x_0^*.

(iv) $\lim_{n\to\infty} \|x_n - T(x_n)\| = r_C(T) := \inf_{x\in C} \|x - T(x)\|$ if $\{\lambda_n\}$ satisfies the conditions:

$$\limsup_{n\to\infty} \lambda_n < 1 \quad \text{and} \quad \sum_{n=0}^{\infty} \lambda_n = \infty.$$

(v) Assume $\sum_{n=0}^{\infty} \lambda_n = \infty$. Then we have

$$\lim_{n\to\infty} \frac{\|x_{n+1} - x_n\|}{\lambda_n} = \lim_{n\to\infty} \frac{\|x_0 - x_n\|}{\sum_{k=0}^{n-1} \lambda_k} = r_C(T).$$

In 1975, Reich [177] established a far-reaching extension of Theorem 3.1 in a very simple and smart manner for nonexpansive mappings.

Theorem 3.6 *Let C be a nonempty closed convex subset of a uniformly convex Banach space X which is Fréchet differentiable. Let $T : C \to C$ be a nonexpansive operator with a fixed point. Let $\{\lambda_n\}$ be a real sequence in $[0, 1]$ such that*

$$\sum_{n=0}^{+\infty} \lambda_n(1 - \lambda_n) = +\infty.$$

Then the sequence $\{x_n\}$ generated by the Krasnosel'skiĭ–Mann iteration (3.1) converges weakly to a fixed point of T.

Note that the assumption that X is Fréchet differentiable in Theorem 3.6 can be replaced with that X satisfies the following so-called *Opial property* [163]. Recall that a Banach space X is said to satisfy Opial's property if for any sequence $\{x_n\}$ in X the condition $x_n \rightharpoonup x^*$ implies

$$\limsup_{n\to\infty} \|x_n - x^*\| < \limsup_{n\to\infty} \|x_n - x\| \quad \text{for all } x \neq x^*.$$

Very recently, a strong convergence result for the Krasnosel'skiĭ–Mann iteration was presented in a Banach space.

Theorem 3.7 ([195, Theorem 1]) *Let X be a real uniformly convex Banach space, C be a nonempty closed convex subset of X and $T : C \to C$ be a nonexpansive mapping with $Fix(T) \neq \emptyset$. Suppose that $S := \mathrm{Id} - T$ is φ-accretive and $\{\lambda_n\}$ satisfies the following condition:*

$$\sum_{n=0}^{+\infty} \lambda_n(1 - \lambda_n) = +\infty.$$

Then the sequence $\{x_n\}$ generated by the Krasnosel'skiĭ–Mann iteration (3.1) converges in norm to a fixed point of T.

The uniform convexity in Theorem 3.7 can be weakened to the strict convexity together with the Kadec-Klee property (see its definition in [195]), the demiclosedness of $S = \mathrm{Id} - T$ and $\lambda_n \le c$ (for some $c \in (0, 1)$), $\sum_{n=0}^{+\infty} \lambda_n = +\infty$.

Xu [196] proposed the following open problem for the convergence of the Krasnosel'skiĭ–Mann iteration in a Banach space:

Open Problem *Let X be a uniformly convex Banach space, C be a nonempty closed bounded convex subset of X, and $T : C \to C$ be a nonexpansive mapping with $\mathrm{Fix}(T) \ne \emptyset$. Let $\{x_n\}$ be generated by the Krasnosel'skiĭ–Mann iteration (3.1). Assume that $\{\lambda_n\} \subset [0, 1]$ such that*

$$\sum_{n=0}^{+\infty} \lambda_n (1 - \lambda_n) = +\infty.$$

Does the sequence $\{x_n\}$ generated by the Krasnosel'skiĭ–Mann iteration (3.1) converge to a fixed point of T?

Recently, Bot and Csetnek [22] obtained the Krasnosel'skiĭ–Mann iteration through a dynamical system associated with the fixed points set of a nonexpansive operator.

Let $T : \mathscr{H} \to \mathscr{H}$ be a nonexpansive operator, $\lambda : [0, +\infty) \to [0, 1]$ be a Lebesgue measurable function and $x_0 \in \mathscr{H}$. They considered the following dynamical system governed by the nonexpansive operator T:

$$\begin{cases} \dot{x}(t) = \lambda(t)(T(x(t)) - x(t)), \\ x(0) = x_0. \end{cases} \tag{3.10}$$

The explicit discretization of (3.10) with respect to the time variable t with step size $h_n > 0$ yields the following iterative scheme:

$$x_{n+1} = x_n + h_n \lambda_n (T(x_n) - x_n) \quad \text{for each } n \ge 0.$$

By taking $h_n \equiv 1$, this becomes the Krasnosel'skiĭ–Mann iteration (3.1).

The weak convergence of the orbits of the dynamic system (3.10) to a fixed point of the operator T is investigated by relying on Lyapunov analysis (see [22, Theorem 6]).

Theorem 3.8 *Let $T : \mathscr{H} \to \mathscr{H}$ be a nonexpansive mapping with $\mathrm{Fix}T \ne \emptyset$, $\lambda : [0, +\infty) \to [0, 1]$ be a Lebesgue measurable function and $x_0 \in \mathscr{H}$. Suppose that one of the following conditions is fulfilled:*

$$\int_0^{+\infty} \lambda(t)(1 - \lambda(t))dt = +\infty \quad \text{or} \quad \inf_{t \ge 0} \lambda(t) > 0.$$

Let $x : [0, +\infty) \to \mathscr{H}$ be the unique strong global solution of the dynamic system (3.10). Then the following statements are true:

(i) $\lim_{t \to +\infty} (T(x(t)) - x(t)) = 0$,

(ii) $x(t)$ converges weakly to a point in FixT as $t \to +\infty$.

The convergence rate of the fixed point residual function $t \mapsto \|T(x(t)) - x(t)\|$ is shown to be $o(\frac{1}{\sqrt{t}})$, i.e.,

$$\|T(x(t)) - x(t)\| \leq \frac{d(x_0, \text{Fix}T)}{\sqrt{\underline{\tau} t}},$$

where $\underline{\tau} = \inf_{t \geq 0} \lambda(t)(1 - \lambda(t)) > 0$.

3.3 The Krasnosel'skiĭ–Mann Iteration with Perturbations

It is important to investigate some perturbations of the Krasnosel'skiĭ–Mann iteration since there exist the errors in the computation of $T(x_n)$ for each $n \geq 0$ and one can construct the inertial type modifications or apply the superiorization idea by using perturbations.

Dong et al. [65] considered the Krasnosel'skiĭ–Mann iteration with perturbations as follows:

$$x_{n+1} = (1 - \lambda_n)(x_n + e_n^1) + \lambda_n T(x_n + e_n^2) \quad \text{for each } n \geq 0. \tag{3.11}$$

The sequences $\{e_n^i\}$ for each $i = 1, 2$ of perturbations are assumed to be summable, i.e.,

$$\sum_{n=0}^{\infty} \|e_n^i\| < +\infty \quad \text{fo each } i = 1, 2. \tag{3.12}$$

Lemma 3.1 Let $T : \mathscr{H} \to \mathscr{H}$ be a nonexpansive operator and the sequence $\{x_n\}$ be generated by (3.11)..Then we have

$$\|x_{n+1} - T(x_{n+1})\| \leq \|x_n - T(x_n)\| + 2(\|e_n^1\| + \|e_n^2\|) \quad \text{for each } n \geq 0. \tag{3.13}$$

Proof From the definition of the sequence $\{x_n\}$, we have

$$x_{n+1} - T(x_{n+1}) = (1 - \lambda_n)(x_n + e_n^1) + \lambda_n T(x_n + e_n^2)$$

$$- T\left[(1 - \lambda_n)(x_n + e_n^1) + \lambda_n T(x_n + e_n^2)\right]$$

$$= (1 - \lambda_n)\left[(x_n + e_n^1) - T(x_n + e_n^2)\right]$$

$$+ T(x_n + e_n^2) - T\left[(1 - \lambda_n)(x_n + e_n^1) + \lambda_n T(x_n + e_n^2)\right],$$

which, with the nonexpansivity of T, implies

$$\|x_{n+1} - T(x_{n+1})\| \leq (1 - \lambda_n)\|(x_n + e_n^1) - T(x_n + e_n^2)\|$$

$$+ \|T(x_n + e_n^2) - T[(1 - \lambda_n)(x_n + e_n^1) + \lambda_n T(x_n + e_n^2)]\|$$

$$\leq (1 - \lambda_n)\|(x_n + e_n^1) - T(x_n + e_n^2)\|$$

$$+ \lambda_n \|(x_n + e_n^1) - T(x_n + e_n^2)\| + \|e_n^1\| + \|e_n^2\|$$

$$= \|(x_n + e_n^1) - T(x_n + e_n^2)\| + \|e_n^1\| + \|e_n^2\|.$$

$$(3.14)$$

Using the nonexpansivity of T again, we have

$$\|(x_n + e_n^1) - T(x_n + e_n^2)\| \leq \|x_n - T(x_n)\| + \|e_n^1\| + \|T(x_n) - T(x_n + e_n^2)\|$$

$$\leq \|x_n - T(x_n)\| + \|e_n^1\| + \|e_n^2\|.$$

$$(3.15)$$

Therefore, combining (3.14) and (3.15), we obtain (3.13). This completes the proof.

Now, we give the convergence of the Krasnosel'skiĭ–Mann iteration with perturbations (3.11).

Theorem 3.9 *Let* $T : \mathscr{H} \to \mathscr{H}$ *be a nonexpansive operator with* $\mathrm{Fix}(T) \neq \emptyset$. *Assume that*

$$\sum_{n=0}^{\infty} \lambda_n(1 - \lambda_n) = \infty.$$

Then the sequence $\{x_n\}$ *generated by* (3.11) *weakly converges to a point in* $\mathrm{Fix}(T)$.

Proof For any $p \in \mathrm{Fix}(T)$, we have

$$\|x_{n+1} - p\|^2 = \|(1 - \lambda_n)(x_n + e_n^1) + \lambda_n T(x_n + e_n^2) - p\|^2$$

$$\leq (1 - \lambda_n)\|x_n + e_n^1 - p\|^2 + \lambda_n\|T(x_n + e_n^2) - p\|^2$$

$$- \lambda_n(1 - \lambda_n)\|(x_n + e_n^1) - T(x_n + e_n^2)\|^2 \qquad (3.16)$$

$$\leq (1 - \lambda_n)\|x_n + e_n^1 - p\|^2 + \lambda_n\|x_n + e_n^2 - p\|^2$$

$$- \lambda_n(1 - \lambda_n)\|(x_n + e_n^1) - T(x_n + e_n^2)\|^2.$$

Using Lemma 2.9, we have

$$(1 - \lambda_n)\|x_n + e_n^1 - p\|^2 + \lambda_n\|x_n + e_n^2 - p\|^2$$

$$\leq (1 - \lambda_n)\left[(1 + \|e_n^1\|)\|x_n - p\|^2 + \|e_n^1\| + \|e_n^1\|^2\right]$$

$$+ \lambda_n\left[(1 + \|e_n^2\|)\|x_n - p\|^2 + \|e_n^2\| + \|e_n^2\|^2\right] \qquad (3.17)$$

$$\leq (1 + \|e_n^1\| + \|e_n^2\|)\|x_n - p\|^2 + \|e_n^1\| + \|e_n^2\| + \|e_n^1\|^2 + \|e_n^2\|^2.$$

Combining (3.16) and (3.17), we have

$$\|x_{n+1} - p\|^2 \leq (1 + \|e_n^1\| + \|e_n^2\|)\|x_n - p\|^2 + \|e_n^1\| + \|e_n^2\| + \|e_n^1\|^2 + \|e_n^2\|^2.$$

Using Lemma 2.8 and the condition (3.12), it follows that $\lim_{n\to\infty}\|x_n - p\|$ exists and hence $\{x_n\}$ is bounded. From (3.16) and (3.17), we have

$$\lambda_n(1 - \lambda_n)\|(x_n + e_n^1) - T(x_n + e_n^2)\|^2$$

$$\leq \|x_n - p\|^2 - \|x_{n+1} - p\|^2 + (\|e_n^1\| + \|e_n^2\|)\|x_n - p\|^2$$

$$+ \|e_n^1\| + \|e_n^2\| + \|e_n^1\|^2 + \|e_n^2\|^2,$$

which yields

$$\sum_{n=0}^{+\infty} \lambda_n(1 - \lambda_n)\|(x_n + e_n^1) - T(x_n + e_n^2)\|^2 < +\infty. \qquad (3.18)$$

Due to $\sum_{n=0}^{\infty} \lambda_n(1 - \lambda_n) = \infty$, we have

$$\liminf_{n\to\infty} \|(x_n + e_n^1) - T(x_n + e_n^2)\| = 0.$$

Since T is nonexpansive, it follows that

$$\|x_n - T(x_n)\| \le \|(x_n + e_n^1) - T(x_n + e_n^2)\| + \|e_n^1\| + \|T(x_n + e_n^2) - T(x_n)\|$$

$$\le \|(x_n + e_n^1) - T(x_n + e_n^2)\| + \|e_n^1\| + \|e_n^2\|.$$

Thus we obtain

$$\liminf_{n \to \infty} \|x_n - T(x_n)\| = 0.$$

Combining Lemmas 2.8, 3.1 and (3.12), we can show that $\lim_{n \to \infty} \|x_n - T(x_n)\|$ exists. So, we claim

$$\lim_{n \to \infty} \|x_n - T(x_n)\| = 0. \tag{3.19}$$

Using the boundedness of $\{x_n\}$, we have $\omega_w(x_n) \ne \emptyset$. By Lemma 2.6 and (3.19), we have $\omega_w(x_n) \subseteq \text{Fix}(T)$. Therefore, applying Lemma 2.7, it follows that $\{x_n\}$ weakly converges a point in $\text{Fix}(T)$. This completes the proof.

Next, we discuss the bounded perturbation resilience of the Krasnosel'skiĭ–Mann iteration, which enable to apply the superiorization idea (see, for example, [31, 32, 101]).

First, we introduce the definition of the bounded perturbation resilience. To do so, we start by introducing the term *"basic algorithm"*.

Let $\Theta \subseteq \mathcal{H}$ and consider the algorithmic operator $\mathbf{A}_\Psi : \Theta \to \mathcal{H}$ which works iteratively by, for any arbitrary starting point $x_0 \in \Theta$,

$$x_{n+1} = \mathbf{A}_\Psi(x_n) \quad \text{for each } n \ge 0, \tag{3.20}$$

which is denoted as the *basic algorithm*. The bounded perturbation resilience (BPR) of the basic algorithm is defined as follows:

Definition 3.2 (Bounded Perturbation Resilience (Shortly, BPR)) An algorithmic operator $\mathbf{A}_\Psi : \Theta \to \mathcal{H}$ is said to be *bounded perturbations resilient* if the algorithm (3.20) generates a sequence $\{x_n\}$ with $x_0 \in \Theta$ converging to a point in Ψ which is superior with respect to the solution set, then a sequence $\{y_n\}$, starting from any $y_0 \in \Theta$, generated by

$$y_{n+1} = \mathbf{A}_\Psi(y_n + \rho_n v_n) \quad \text{for each } n \ge 0 \tag{3.21}$$

also converges to a point in Ψ, provided that

(a) the sequence $\{v_n\}_{n \in \mathbb{N}}$ is bounded;
(b) $\{\rho_n\} \subset \mathbb{R}$ is positive and satisfies $\sum_{n=0}^{\infty} \rho_n < \infty$;
(c) $y_n + \rho_n v_n \in \Theta$ for each $n \ge 0$.

The condition (c) is needed only if $\Theta \ne \mathcal{H}$, in which (c) is enforced in the superiorized version of the basic algorithm (see Step (14) in the *"Superiorized*

Version of Algorithm P" in [102, pp. 5537] and Step (14) in *"Superiorized Version of the ML-EM Algorithm"* in [84, Subsection II.B]). This will be the case in the present work.

Treating the Krasnosel'skiĭ–Mann iteration as the basic algorithm $\mathbf{A}_\psi$, our strategy is to prove the convergence of the basic algorithms with bounded perturbations and then show how this yields the BPR of the algorithms according to Definition 3.2.

A superiorized version of any basic algorithm employs the perturbed version of the basic algorithm as in (3.21). A certificate to do so in the superiorization method (see [29]) is gained by showing that the basic algorithm is BPR. Therefore, proving the BPR of an algorithm is the first step toward superiorizing it.

To this end, first, we treat the right-hand side of the Krasnosel'skiĭ–Mann iteration (3.1) as the algorithmic operator $\mathbf{A}_\psi$ of Definition 3.2, that is, we define as follows: for each $n \geq 0$,

$$\mathbf{A}_\psi^n = (1 - \lambda_n)\mathrm{Id} + \lambda_n T.$$

According to Definition 3.2, the *bounded perturbation* of the Krasnosel'skiĭ–Mann iteration is defined as follows: for each $n \geq 0$,

$$\begin{aligned}
x_{n+1} &= \mathbf{A}_\psi^n (x_n + \rho_k v_n) \\
&= (1 - \lambda_n)(x_n + \rho_n v_n) + \lambda_n T(x_n + \rho_n v_n).
\end{aligned} \tag{3.22}$$

Setting $e_n^1 = e_n^2 = \rho_n v_n$ in the scheme (3.11), then it becomes the above scheme (3.22).

Assume that the sequence $\{v_k\}$ in $\mathscr{H}$ is bounded and the scalar sequence $\{\rho_n\}$ is positive for each $n \geq 0$ and satisfy the condition:

$$\sum_{n=0}^{\infty} \rho_n < \infty.$$

Using Theorem 3.9, we prove the convergence of the scheme (3.22).

Theorem 3.10 *Let $T : \mathscr{H} \to \mathscr{H}$ be a nonexpansive operator with $\mathrm{Fix}(T) \neq \emptyset$. Assume that*

$$\sum_{n=0}^{\infty} \lambda_n(1 - \lambda_n) = \infty.$$

Then the sequence $\{x_n\}$ generated by (3.22) weakly converges to a point in $\mathrm{Fix}(T)$.

Remark 3.1

(1) From Definition 3.2 and Theorem 3.9, it follows that the Krasnosel'skiĭ–Mann iteration for nonexpansive operators is bounded perturbations resilient.

(2) Based on the bounded perturbation resilience of an algorithm, one can construct
    the inertial type extrapolation of the algorithm. By using this idea, the authors
    [60] introduced some inertial type projection and contraction algorithms.

3.4   The Convergence Rate

The convergence rate is one of the main criterions to evaluate algorithms. In this
section, we present several convergence rates of the Krasnosel'skiĭ–Mann iteration.

Let $T : \mathscr{H} \to \mathscr{H}$ be L-Lipschitz and $-T$ be monotone. He et al. [99] presented
an estimation of the degree of the convergence of the Krasnosel'skiĭ–Mann iteration.

Theorem 3.11 *Assume that* $\lambda_n \equiv \lambda$ *satisfies* $\max\{\frac{L^2-1}{L^2+1}, 0\} < \lambda < 1$. *Let* $\{x_n\}$ *be
generated by the Krasnosel'skiĭ–Mann iteration* (3.1). *Then we have the estimation
of degree of convergence:*

$$\|x_n - x^*\| \leq \frac{[\lambda^2 + (1-\lambda)^2 L^2]^{\frac{n}{2}}}{1 - \sqrt{\lambda^2 + (1-\lambda)^2 L^2}} \|x_1 - x_0\|,$$

where x^* *is a unique fixed point of* T. *In addition, when* $\lambda = \frac{L^2}{L^2+1}$ *(the optimal
parameter), the following estimation of degree of convergence is optimal in the sense
of ignoring constant factors:*

$$\|x_n - x^*\| \leq \sqrt{L^2 + 1}(1 + \sqrt{L^2 + 1}) \left(\frac{L}{\sqrt{L^2+1}}\right)^n \|x_1 - x_0\|.$$

When T is a general operator, such as nonexpansive operator, it is difficult to
obtain the estimation of the distance of the iterative sequence $\{x_n\}$ and a fixed point
(or fixed point set).

Recall that the crucial step in proving the convergence of the Krasnosel'skiĭ–
Mann iteration is to show that the fixed point residuals $\|x_n - T(x_n)\|$ tend to 0. This
property is also known as the asymptotic regularity [12, 21]. The asymptotic regularity of the Krasnosel'skiĭ–Mann iteration was proved under various assumptions
(see, for example, [89]).

Goebel and Kirk [87] showed that, for each $\epsilon > 0$, it holds $\|x_n - T(x_n)\| \leq \epsilon$ for
each $n \geq n_0$ with n_0 depending on ϵ and C, but independent of the initial point x_0
and the mapping T. Recently, it was proved that n_0 could be chosen to depend on C
only through its diameter in [118, 119].

Some authors have investigated the convergence rate of the Krasnosel'skiĭ–Mann
iteration by using the fixed point residuals $\|x_n - T(x_n)\|$ in the last twenty years.

Let X be a Banach space, C be a nonempty closed convex bounded subset of X,
and $T : C \to C$ be a nonexpansive mapping. Baillon and Bruck [11] proved that

$$\|x_n - T(x_n)\| = O(1/\log(n))$$

for $\lambda_n \equiv \lambda$ for each $n \geq 0$ and conjectured the existence of a universal constant κ such that

$$\|x_n - T(x_n)\| \leq \kappa \frac{\text{diam}(C)}{\sqrt{\sum_{i=0}^{n} \lambda_i(1 - \lambda_i)}}. \tag{3.23}$$

When T is an affine mapping, they gave a tight bound with $\kappa = 1/\sqrt{2\pi}$ for $\lambda_n \equiv 1/2$ and showed that, for the general relaxation parameter λ_n, the optimal constant is $\kappa = \max_{x \geq 0} \sqrt{x}e^{-x}I_0(x) \sim 0.4688$, where $I_0(\cdot)$ is a modified Bessel function of the first kind.

The conjecture (3.23) was proved in [12] for the case $\lambda_n \equiv \lambda$ for each $n \geq 0$ with $\kappa = 1/\sqrt{\pi}$. For any sequence $\{\lambda_n\}$, the conjecture (3.23) was settled by using the connection between the iterates and a stochastic process involving sums of non-homogeneous Bernoulli trials in [51].

Recently, Bravo and Cominetti [26] showed that $\kappa = 1/\sqrt{\pi}$ in (3.23) is tight, that is, the optimal constant of asymptotic regularity is exactly $1/\sqrt{\pi}$.

Theorem 3.12 ([26, Theorem 1.1]) *The constant $\kappa = 1/\sqrt{\pi}$ in the bound (3.23) is tight. Specifically, for each $\kappa < 1/\sqrt{\pi}$, there exists a nonexpansive mapping T defined on the unit cube $C = [0, 1]^{\mathbb{N}} \subseteq \ell^{\infty}(\mathbb{N})$, an initial point $x_0 \in C$, and a constant sequence $\lambda_n \equiv \lambda$ such that the corresponding Krasnosel'skiĭ–Mann iterates $\{x_n\}$ satisfies the condition: for some $n \geq 0$,*

$$\|x_n - T(x_n)\| > \kappa \frac{\text{diam}(C)}{\sqrt{\sum_{i=0}^{n} \lambda_i(1 - \lambda_i)}}. \tag{3.24}$$

Matsushita [148] established a convergence rate which improves big-O rate to little-o rate without any additional restrictions.

Theorem 3.13 ([148, Theorem 3.1]) *Let C be a nonempty closed convex subset of a Hilbert space $\mathcal{H}$, $T : C \to C$ be a nonexpansive mapping such that $\text{Fix}(T) \neq \emptyset$. Let $\{x_n\}$ be the sequence generated by the Krasnosel'skiĭ–Mann iteration (3.1), where $x_0 \in C$, and $\{\lambda_n\}$ is a sequence in $[0, 1]$ such that*

$$\lim_{n \to \infty} \sigma_n = \infty,$$

where $\sigma_n := \sum_{j=0}^{n} \lambda_j(1 - \lambda_j)$, $\forall n \geq 0$. Then the convergence rate estimate

$$\|x_n - T(x_n)\| = o(1/\sqrt{\sigma_n})$$

holds, that is, $\lim_{n \to \infty} \sqrt{\sigma_n}\|x_n - T(x_n)\| = 0$.

The proof of Theorem 3.13 is very simple and relies on the connection between the Krasnosel'skiĭ–Mann iteration and a useful technique on the convergence rate of summable sequences.

Liang et al. [126] presented the global pointwise and ergodic iteration-complexity bounds for the following inexact Krasnosel'skiĭ–Mann iteration:

$$x_{n+1} = (1 - \lambda_n)x_n + \lambda_n(T(x_n) + e_n) \quad \text{for each } n \geq 1, \tag{3.25}$$

where e_n is the error of approximating $T(x_n)$.

Recently, Dong et al. [65] considered the complexity bounds for the Krasnosel'skiĭ–Mann iteration with perturbations (3.11), which include the results in [126] as special cases.

Define

$$\overline{\tau} = \sup_{n \in \mathbb{N}} \lambda_n(1 - \lambda_n), \quad \underline{\tau} = \inf_{n \in \mathbb{N}} \lambda_n(1 - \lambda_n),$$

and

$$d_1 = d(x_1, \text{Fix}(T)) = \inf_{x \in \text{Fix}(T)} \|x_1 - x\|.$$

Set

$$\upsilon_1 = 4 \sup_{n \in \mathbb{N}} \{\|x_n - T(x_n)\| + \|e_n^1\| + \|e_n^2\|\},$$

$$\upsilon_2 = \sup_{n \in \mathbb{N}} \{2\|(x_n + e_n^1) - T(x_n + e_n^2)\| + \|e_n^1\| + \|e_n^2\|\}$$

and

$$\upsilon_3 = \sup_{n \in \mathbb{N}} \{2\|x_n - p\| + (1 - \lambda_n)\|e_n^1\| + \lambda_n\|e_n^2\|\},$$

where $p \in \text{Fix}(T)$.

Theorem 3.14 (Pointwise Iteration-Complexity Bound) *For the Krasnosel'skiĭ–Mann iteration with perturbations, if there holds*

$$0 < \inf_{n \in \mathbb{N}} \lambda_n \leq \sup_{n \in \mathbb{N}} \lambda_n < 1 \quad and \quad \left\{(n + 1)(\|e_n^1\| + \|e_n^2\|)\right\} \in \ell_+^1, \tag{3.26}$$

then, denoting

$$C_1 = (\upsilon_2 + \upsilon_3) \sum_{j \in \mathbb{N}} (\|e_j^1\| + \|e_j^2\|) + \upsilon_1 \overline{\tau} \sum_{\ell \in \mathbb{N}} (\ell + 1)(\|e_\ell^1\| + \|e_\ell^2\|) < \infty,$$

we have

$$\|x_n - T(x_n)\| \leq \sqrt{\frac{d_1^2 + C_1}{\underline{\tau}(n+1)}}. \tag{3.27}$$

Proof From (3.13), we have

$$\|x_{n+1} - T(x_{n+1})\|^2 \leq \|x_n - T(x_n)\|^2 + \upsilon_1(\|e_n^1\| + \|e_n^2\|). \tag{3.28}$$

By using the Cauchy–Schwarz inequality, we get

$$\|x_n - T(x_n)\| \leq \|(x_n + e_n^1) - T(x_n + e_n^2)\| + \|e_n^1\| + \|T(x_n) - T(x_n + e_n^2)\|$$

$$\leq \|(x_n + e_n^1) - T(x_n + e_n^2)\| + \|e_n^1\| + \|e_n^2\|,$$

which implies that

$$\|x_n - T(x_n)\|^2 \leq \|(x_n + e_n^1) - T(x_n + e_n^2)\|^2 + \upsilon_2(\|e_n^1\| + \|e_n^2\|). \tag{3.29}$$

Similarly, we easily get

$$(1-\lambda_n)\|x_n+e_n^1-p\|^2+\lambda_n\|x_n+e_n^2-p\|^2 \leq \|x_n-p\|^2+\upsilon_3(\|e_n^1\|+\|e_n^2\|). \tag{3.30}$$

Combining (3.16), (3.29), and (3.30), we obtain

$$\|x_{n+1} - p\|^2$$

$$\leq \|x_n - p\|^2 - \lambda_n(1-\lambda_n)\|x_n - T(x_n)\|^2 + (\upsilon_2 + \upsilon_3)(\|e_n^1\| + \|e_n^2\|). \tag{3.31}$$

Using (3.28), (3.31) and employing the proof of [126, Theorem 1], we get (3.27). This completes the proof. ∎

To give the ergodic iteration-complexity bound of (3.11), define $\Lambda_n = \sum_{j=0}^n \lambda_j$.

Theorem 3.15 (Ergodic Iteration-Complexity Bound) *Suppose that*

$$C_2 = \sum_{n=0}^{\infty}[(1 - \lambda_n)\|e_n^1\| + \lambda_n\|e_n^2\|] < \infty.$$

Then we have

$$\left\|\frac{1}{\Lambda_n}\sum_{j=0}^n \lambda_j(x_j - T(x_j))\right\| \leq \frac{2(d_1 + C_2)}{\Lambda_n}. \tag{3.32}$$

In particular, if $\inf_{n\in\mathbb{N}} \lambda_n > 0$, *then* $\left\|\frac{1}{\Lambda_n}\sum_{j=0}^n \lambda_j(x_j - T(x_j))\right\| = O(1/n)$.

Proof Again, let $p \in \text{Fix}(T)$ such that $d_1 = \|x_1 - p\|$. Since T is nonexpansive, we have

$$\|x_{n+1} - p\| \leq (1 - \lambda_n)\|x_n + e_n^1 - p\| + \lambda_n\|T(x_n + e_n^2) - p\|$$
$$\leq \|x_n - p\| + (1 - \lambda_n)\|e_n^1\| + \lambda_n\|e_n^2\|$$
$$\leq \cdots \tag{3.33}$$
$$\leq \|x_1 - p\| + \sum_{j=0}^{n}\left[(1 - \lambda_j)\|e_j^1\| + \lambda_j\|e_j^2\|\right].$$

Using this inequality and (3.11), we get

$$\left\|\frac{1}{\Lambda_n}\sum_{j=0}^{n}\lambda_j(x_j - T(x_j))\right\|$$
$$= \frac{1}{\Lambda_n}\left\|\sum_{j=0}^{n}(x_j - x_{j+1}) + \sum_{j=0}^{n}(1 - \lambda_j)e_j^1 + \sum_{j=0}^{n}\lambda_j(T(x_j + e_j^2) - T(x_j))\right\|$$
$$\leq \frac{1}{\Lambda_n}\left(\|x_1 - p\| + \|x_{n+1} - p\| + \sum_{j=0}^{n}\left[(1 - \lambda_j)\|e_j^1\| + \lambda_j\|e_j^2\|\right]\right)$$
$$\leq \frac{2(d_1 + C_2)}{\Lambda_n}.$$

Thus we have the conclusion. This completes the proof.

Recently, Bravo et al. [27] established a metric bound for the fixed point residuals of the inexact Krasnosel'skiĭ–Mann iteration (3.25) and derived the rate of convergence as well as the convergence of the iterates towards a fixed point in Banach spaces.

Let

$$\tau_n = \sum_{k=0}^{n}\lambda_k(1 - \lambda_k)$$

and define a function $\sigma : [0, \infty) \to \mathbb{R}$ by

$$\sigma(y) = \min\left\{1, \frac{1}{\sqrt{\pi y}}\right\}.$$

Theorem 3.16 ([27, Theorem 3]) *Let X be a Banach space, C be a closed convex subset of X, and $T : C \to C$ be a nonexpansive mapping. Let $\{x_n\}$ be the sequence generated by the scheme (3.25) and assume that there exists $\kappa \geq 0$ such that*

$$x_n \in C \ and \ \|T(x_n) - x_0\| \leq \kappa \ for \ each \ n \geq 0. \tag{3.34}$$

Suppose that $\sum_{n=0}^{+\infty} \|e_n\| < +\infty$ and that λ_n is bounded away from 0 and 1. Then there exists a constant $c \geq 0$ such that

$$\|x_n - T(x_n)\| \leq \frac{c}{\sqrt{n}} + \sum_{i \geq \lfloor \frac{n}{2} \rfloor} 2\|e_i\|.$$

Moreover, if $\varphi : [0, \infty) \to [0, \infty)$ is nondecreasing and $\mu = \sum_{n=0}^{\infty} \varphi(n)\|e_n\| < \infty$, then

$$\|x_n - T(x_n)\| \leq \frac{c}{\sqrt{n}} + \frac{2\mu}{\varphi(\lfloor \frac{n}{2} \rfloor)}.$$

In particular, if $\sum_{n=0}^{\infty} n^a \|e_n\| < \infty$ for some $a \geq 0$ then $\|x_n - T(x_n)\| = O(1/n^b)$ with $b = \min\{\frac{1}{2}, a\}$.

In fact, the convergence rate depends on the error (see [27, Corollary 2]). Suppose $\|e_n\| = O(1/n^a)$ with $a \geq \frac{1}{2}$ and λ_n is bounded away from 0 and 1. Then we have the following:

(i) If $\frac{1}{2} \leq a < 1$, then $\|x_n - T(x_n)\| = O(1/n^{a-1/2})$.
(ii) If $a = 1$, then $\|x_n - T(x_n)\| = O(\log n/\sqrt{n})$.
(iii) If $a > 1$, then $\|x_n - T(x_n)\| = O(1/\sqrt{n})$.

Until now, the researchers only obtained sublinear convergence rates of the Krasnosel'skiĭ–Mann iteration for nonexpansive mappings, which means that the iterative sequence achieves initially quick progress towards some optimal point and subsequently levels off.

The Krasnosel'skiĭ–Mann iteration is typical of iteration methods involving nonexpansive mappings and its convergence may be arbitrarily slow. Actually, Oblomskaja [161] gave a linear example where the convergence is slower than $n^{-\tau}$ for all $\tau \in (0, 1)$. Some ways to remedy this behavior have been proposed, mainly including relaxation strategies and the inertial acceleration (see, for example, [114]), which will be discussed in Chaps. 5–7.

Chapter 4
Relations of the Krasnosel'skiĭ–Mann Iteration and the Operator Splitting Methods

Some operator splitting methods can be treated as particular cases of the Krasnosel'skiĭ–Mann iteration for finding fixed points of nonexpansive operators or averaged operators. Since the Krasnosel'skiĭ–Mann iteration has the advantage of allowing the inclusion of relaxation parameters in the update rules of the iterates, it is often used in the design and analysis of operator splitting methods.

In this chapter, we provide a brief overview of several classical operator splitting methods which can be seen as special cases of the Krasnosel'skiĭ–Mann iteration (3.1). These methods include the gradient descent algorithm, the proximal point algorithm, the forward-backward splitting method, the backward-forward splitting method, the Douglas–Rachford splitting method, the Davis–Yin splitting method, and several primal-dual splitting methods.

4.1 The Gradient Descent Algorithm

Let $f : \mathscr{H} \to \mathbb{R}$ be a convex function with L-Lipschitz gradient. Consider the following *smooth convex minimization*:

$$\min_{x \in \mathscr{H}} f(x). \tag{4.1}$$

A simple and classical method for approximating (4.1) is the gradient descent algorithm [172].

Gradient Descent Algorithm Let $f : \mathscr{H} \to \mathbb{R}$ be a convex function with L-Lipschitz gradient and $\{\gamma_n\} \in (0, 2/L)$. Choose $x_0 \in \mathscr{H}$ arbitrarily and compute x_{n+1} based on the following rule:

$$x_{n+1} = x_n - \gamma_n \nabla f(x_n) \quad \text{for each } n \geq 0. \tag{4.2}$$

© The Author(s), under exclusive license to Springer Nature Switzerland AG 2022
Q.-L. Dong et al., *The Krasnosel'skiĭ-Mann Iterative Method*, SpringerBriefs in
Optimization, https://doi.org/10.1007/978-3-030-91654-1_4

From Baillon–Haddad's Theorem, we have

$$\nabla f \text{ is } L\text{-Lipschitz} \iff T := \text{Id} - \frac{2}{L}\nabla f \text{ is nonexpansive.}$$

Furthermore, $\text{Argmin}_{x\in\mathcal{H}} f(x) = \text{Fix}(T)$.

Therefore, the scheme (4.2) can be seen as the Krasnosel'skiĭ–Mann iteration

$$x_{n+1} = (1 - \lambda_n)\, x_n + \lambda_n T x_n \quad \text{for each } n \geq 0,$$

where $\lambda_n = \frac{\gamma_n L}{2}$.

4.2 The Proximal Point Algorithm

Fermat's theorem 2.1 indicates that minimizing a function $f \in \Gamma_0(\mathcal{H})$ is equivalent to finding the zeros of its subdifferential operator ∂f. Therefore, a minimizer of f is a fixed point of $J_{\partial f}$.

In general, let A be a maximally monotone operator such that $\text{zer}(A) \neq \emptyset$. The following problem:

$$\text{Find } x \in \mathcal{H} \text{ such that } 0 \in A(x), \tag{4.3}$$

i.e., finding an element of $\text{zer}(A)$, is called the *monotone inclusion problem*.

A fundamental algorithm for solving the problem (4.3) is the proximal point algorithm [147, 178]. The well-known method of multipliers [103] for constrained convex optimization is known to be a special case of the proximal point algorithm.

Proximal Point Algorithm. Let $A : \mathcal{H} \rightrightarrows \mathcal{H}$ be a maximally monotone operator such that $\text{zer}(A) \neq \emptyset$, $\{\gamma_n\} \in (0, +\infty]$ and $\{\lambda_n\} \in [0, 2]$. Choose $x_0 \in \mathcal{H}$ arbitrarily and compute x_{n+1} based on the following rule:

$$x_{n+1} = (1 - \lambda_n)x_n + \lambda_n J_{\gamma_n A}(x_n) \quad \text{for each } n \geq 0. \tag{4.4}$$

Historically, the scheme (4.4) with $\lambda_n = 1$ for each $n \geq 0$ was named as the *proximal point algorithm* by Rockafellar [178] and the general scheme (4.4) was called the *generalized proximal point algorithm* by Eckstein and Bertsekas [78].

4.3 The Operator Splitting Methods

In practice, we often encounter with problems which have more structures than the problem (4.3), for instance, finding the zeros of the sum of monotone operators.

Problem 4.1 Let $B : \mathcal{H} \to \mathcal{H}$ be β-cocoercive for some $\beta > 0$, $m \geq 1$ be a positive integer and, for each $i \in \{1, \cdots, m\}$, let $A_i : \mathcal{H} \rightrightarrows \mathcal{H}$ be a maximally monotone operator. Consider the following problem:

$$\text{Find } x \in \mathcal{H} \text{ such that } 0 \in B(x) + \sum_{i=1}^{m} A_i(x). \tag{4.5}$$

There are also situations where A_i is composed with linear operators or even parallel sum of maximally monotone operators are involved. In principal, the proximal point algorithm can still be applied to solving such problems, however, even if the resolvent of B and each A_i can be computed, the resolvent of $B + \sum_i A_i$ in most cases is not accessible.

To avoid this difficulty, a wise strategy is to design some iterative schemes such that the resolvents of A_i are computed separately and use the cocoercivity of B. This is the reason that these schemes are called operator splitting methods.

4.3.1 The Forward-Backward Splitting and Backward-Forward Splitting Methods

For Problem 4.1, let $m = 1$.

Problem 4.1.1 Consider finding the zeros of the sum of a maximally monotone operator and a cocoercive operator:

$$\text{Find } x \in \mathcal{H} \text{ such that } 0 \in (A + B)(x), \tag{4.6}$$

where

(a) $A : \mathcal{H} \rightrightarrows \mathcal{H}$ is maximally monotone;
(b) $B : \mathcal{H} \to \mathcal{H}$ is β-cocoercive for some $\beta > 0$;
(c) $\text{zer}(A + B) \neq \emptyset$.

(A) The forward-backward splitting method

Before describing the forward-backward splitting method, we first consider the following averaged operator:

Lemma 4.1 *Let* $A : \mathcal{H} \rightrightarrows \mathcal{H}$ *be a maximally monotone operator and* $B : \mathcal{H} \to \mathcal{H}$ *be a* β-*cocoercive mapping for some* $\beta > 0$. *Choose* $\gamma \in (0, 2\beta)$ *and define*

$$T_{FB} = J_{\gamma A}(\text{Id} - \gamma B).$$

Then we have the following:

(1) T_{FB} *is* $\frac{2\beta}{4\beta - \gamma}$-*averaged.*

(2) $zer(A + B) = Fix(T_{FB})$.

Proof See Lemmas 2.2, 2.11 and [15, Proposition 26.1].

Let $\gamma \in (0, 2\beta)$ and $\{\lambda_n\}$ in $\left[0, \frac{4\beta - \gamma}{2\beta}\right]$ such that

$$\sum_{n=0}^{\infty} \lambda_n \left(\frac{4\beta - \gamma}{2\beta} - \lambda_n\right) = \infty.$$

Choose $x_0 \in \mathcal{H}$ arbitrarily and apply the following iteration:

$$x_{n+1} = (1 - \lambda_n)x_n + \lambda_n T_{FB}x_n \quad \text{for each } n \geq 0. \tag{4.7}$$

The scheme (4.7) is called the *forward-backward splitting method* since it consists of an explicit forward step, followed by an implicit step where the resolvent is computed.

The stepsize γ can be varying along iterations and this result in the non-stationary version of the forward-backward splitting method which is studied in [42, 49]. Problem 4.1.1 with $m \geq 2$ is considered in [175], in which the authors developed a generalized forward-backward splitting method.

(B) The backward-forward splitting method

Before describing the backward-forward splitting method, we first consider the following averaged operator:

Lemma 4.2 *Let $A : \mathcal{H} \rightrightarrows \mathcal{H}$ be a maximally monotone operator and $B : \mathcal{H} \to \mathcal{H}$ be a β-cocoercive mapping for some $\beta > 0$. Choose $\gamma \in (0, 2\beta)$ and define*

$$T_{BF} = (\text{Id} - \gamma B)J_{\gamma A}.$$

Then we have the following:

(1) *T_{BF} is $\frac{2\beta}{4\beta - \gamma}$-averaged.*
(2) *$zer(A_\gamma + B_{-\gamma}) = Fix(T_{BF})$.*

Proof See Lemmas 2.2, 2.11 and [5, Proposition 2.4(ii)].

Let $\gamma \in (0, 2\beta)$ and $\{\lambda_n\}$ in $\left[0, \frac{4\beta - \gamma}{2\beta}\right]$ such that

$$\sum_{n=0}^{\infty} \lambda_n \left(\frac{4\beta - \gamma}{2\beta} - \lambda_n\right) = \infty.$$

Choose $x_0 \in \mathcal{H}$ arbitrarily and apply the following iteration:

$$x_{n+1} = (1 - \lambda_n)x_n + \lambda_n T_{BF}x_n \quad \text{for each } n \geq 0. \tag{4.8}$$

The backward-forward splitting method was introduced by Attouch et al. [5] which arises when studying the time discretization of the regularized Newton method for maximally monotone operators. The backward-forward splitting method has a natural link with the forward-backward splitting method since it involves the same basic blocks, but organized differently.

4.3.2 The Douglas–Rachford Splitting Method

For Problem 4.1, let $B = 0$ and $m = 2$.

Problem 4.1.2 Consider the problem to find the zeros of the sum of two maximally monotone operators:

$$\text{Find } x \in \mathcal{H} \text{ such that } 0 \in (A_1 + A_2)(x), \tag{4.9}$$

where

(a) $A_1, A_2 : \mathcal{H} \rightrightarrows \mathcal{H}$ are maximally monotone;
(b) $\text{zer}(A_1 + A_2) \neq \emptyset$.

The Douglas–Rachford splitting method was originally introduced in [76] to solve nonlinear heat flow problems and later Lions and Mercier [130] extended it to two closed convex sets with nonempty intersection. It is regarded as one of the most popular algorithms to solve (4.9). Recently, the Douglas–Rachford splitting method was extended to nonconvex optimizations and succeeded in solving nonconvex feasibility problems [3, 124]. The relaxation form of the Douglas–Rachford splitting method is considered in [41, 42] and exhibits the advantage in the numerical performance (see, for example, [78, 94]).

Before describing the Douglas–Rachford splitting method, we first consider the following averaged operator:

Lemma 4.3 Let $A_1, A_2 : \mathcal{H} \rightrightarrows \mathcal{H}$ be two maximally monotone operators, $\gamma > 0$ and define

$$T_{DR} := \frac{1}{2}(R_{\gamma A_1} R_{\gamma A_2} + \text{Id}).$$

Then we have the following:

(1) $R_{\gamma A_1} R_{\gamma A_2}$ is nonexpansive.
(2) $T_{DR} = J_{\gamma A_1}(2J_{\gamma A_2} - \text{Id}) - J_{\gamma A_2} + \text{Id}$ is firmly nonexpansive.
(3) $\text{zer}(A_1 + A_2) = J_{\gamma A_2}(\text{Fix}(T_{DR}))$.

The Douglas–Rachford splitting method. For Problem 4.1.2, let $\gamma > 0$ and $\{\lambda_n\}_{n \in \mathbb{N}}$ in $[0, 2]$ such that

$$\sum_{n=0}^{\infty} \lambda_n (2 - \lambda_n) = \infty.$$

Choose $x_0 \in \mathscr{H}$ arbitrarily and apply the following iteration: for each $n \geq 0$,

$$\begin{cases} y_n = J_{\gamma A_2} x_n, \\ z_n = J_{\gamma A_1} (2y_n - x_n), \\ x_{n+1} = x_n + \lambda_n (z_n - y_n). \end{cases} \qquad (4.10)$$

The Douglas–Rachford splitting method (4.10) can be written as the fixed point iteration of the operator T_{DR}, which reads

$$x_{n+1} = (1 - \lambda_n)x_n + \lambda_n T_{DR}(x_n) \quad \text{for each } n \geq 0. \qquad (4.11)$$

When $\lambda_n \equiv 2$, the corresponding scheme is called the *Peaceman–Rachford splitting method* introduced in [166].

By using a product space trick ([184]), the Douglas–Rachford splitting method can be extended to the case $m \geq 2$. The Douglas–Rachford splitting method is a special case of the proximal point algorithm [78]. In the context of convex optimization, the Douglas–Rachford splitting method is closely related to the well-known alternating direction method of multipliers and the primal-dual hybrid gradient method [35, 129, 156, 162].

4.3.3 The Davis–Yin Splitting Method

For Problem 4.1, now let $m = 2$.

Problem 4.1.3 Consider the problem to find the zeros of the sum of two maximally monotone operators and a cocoercive operator:

$$\text{Find } x \in \mathscr{H} \text{ such that } 0 \in (A_1 + A_2 + B)(x), \qquad (4.12)$$

where

(a) $A_1, A_2 : \mathscr{H} \rightrightarrows \mathscr{H}$ are maximally monotone;
(b) $B : \mathscr{H} \to \mathscr{H}$ is β-cocoercive for some $\beta > 0$;
(c) $\text{zer}(A_1 + A_2 + B) \neq \emptyset$.

The Davis–Yin splitting method was recently introduced in [55] to solve the problem (4.12), which includes, as special cases, the forward-backward splitting method, the backward-forward splitting method, and the Douglas–Rachford splitting method.

Before describing the Davis–Yin splitting method, we first consider the following averaged operator:

Lemma 4.4 *Let $A_1, A_2 : \mathscr{H} \rightrightarrows \mathscr{H}$ be two maximally monotone operators and $B : \mathscr{H} \to \mathscr{H}$ be a β-cocoercive mapping for some $\beta > 0$. Choose $\gamma \in (0, 2\beta)$ and define*

$$T_{DY} = J_{\gamma A_1}(2J_{\gamma A_2} - \mathrm{Id} - \gamma B J_{\gamma A_2}) + \mathrm{Id} - J_{\gamma A_2}. \qquad (4.13)$$

Then we have the following:

(1) *T_{DY} is $\frac{2\beta}{4\beta-\gamma}$-averaged. In particular, the following inequality holds: for all $x, y \in \mathscr{H}$,*

$$\|T_{DY}(x) - T_{DY}(y)\|^2 \le \|x - y\|^2 - \frac{2\beta - \gamma}{2\beta}\|(\mathrm{Id} - T_{DY})(x) - (\mathrm{Id} - T_{DY})(y)\|^2.$$

(2) *$zer(A + B) = J_{\gamma A_2}(Fix(T_{DY}))$.*

The Davis–Yin Splitting Method Let $A_1, A_2 : \mathscr{H} \rightrightarrows \mathscr{H}$ be maximally monotone and $B : \mathscr{H} \to \mathscr{H}$ be β-cocoercive for some $\beta > 0$. Let $\gamma \in (0, 2\beta)$ and $\{\lambda_n\}$ in $\left[0, \frac{4\beta-\gamma}{2\beta}\right]$ such that

$$\sum_{n=0}^{\infty} \lambda_n\left(\frac{4\beta - \gamma}{2\beta} - \lambda_n\right) = \infty.$$

Choose $x_0 \in \mathscr{H}$ arbitrarily and apply the following iteration: for each $n \ge 0$,

$$\begin{cases} y_n = J_{\gamma A_2}x_n, \\ z_n = J_{\gamma A_1}(2y_n - x_n - \gamma B y_n), \\ x_{n+1} = x_n + \lambda_n(z_n - y_n). \end{cases} \qquad (4.14)$$

The Davis–Yin splitting method (4.14) can be written as the fixed point iteration of the operator T_{DY}, which reads

$$x_{n+1} = (1 - \lambda_n)x_n + \lambda_n T_{DY}(x_n) \quad \text{for each } n \ge 0. \qquad (4.15)$$

4.3.4 The Primal-Dual Splitting Method

Some special cases of monotone inclusion problem (4.5) are instantiations of the following formulation, which involves an arbitrary number of maximally monotone operators and compositions with linear operator.

Problem 4.1.4 Let $\mathscr{H}, \mathscr{G}$ be two real Hilbert spaces and suppose that

(a) $A : \mathscr{H} \rightrightarrows \mathscr{H}$ is a maximally monotone operator and $B : \mathscr{H} \to \mathscr{H}$ is a β_B-cocoercive mapping for some $\beta_B > 0$;
(b) $C : \mathscr{H} \rightrightarrows \mathscr{H}$ is a maximally monotone operator and D is a β_D-strongly monotone mapping for some $\beta_D > 0$;
(c) $L : \mathscr{H} \to \mathscr{G}$ is a bounded linear operator.

Consider the following *primal monotone inclusion problem*:

$$\text{Find } x \in \mathscr{H} \text{ such that } 0 \in (A + B)(x) + L^*((C \square D)(Lx)) \tag{4.16}$$

and the corresponding *dual problem*:

$$\text{Find } v \in \mathscr{G} \text{ such that } (\exists x \in \mathscr{H}) \begin{cases} 0 \in (A + B)(x) + L^* v, \\ 0 \in (C^{-1} + D^{-1})(v) - Lx. \end{cases} \tag{4.17}$$

(d) The solution sets of the problems (4.16) and (4.17), respectively, denoted by $\mathscr{P}$ and $\mathscr{D}$, are both nonempty.

The study of the primal-dual splitting method dates back to the late 1950s. Since then, various primal-dual splitting methods have been proposed in the literature. Below we introduce the primal-dual splitting method presented in [48], which covers the methods in [35, 52] as special cases.

(A) The primal-dual splitting method

For Problem 4.1.4, choose $\gamma_A, \gamma_C > 0$ such that

$$2 \min\{\beta_B, \beta_D\} \min \left\{ \frac{1}{\gamma_A}, \frac{1}{\gamma_C} \right\} (1 - \sqrt{\gamma_A \gamma_C \|L\|^2}) > 1. \tag{4.18}$$

Choose $x_0 \in \mathscr{H}$ and $v_0 \in \mathscr{G}$ arbitrarily and apply the following iteration: for each $n \geq 0$,

$$\begin{cases} x_{n+1} = J_{\gamma_A A}(x_n - \gamma_A B(x_n) - \gamma_A L^* v_n), \\ \bar{x}_{n+1} = 2x_{n+1} - x_n, \\ v_{n+1} = J_{\gamma_C C^{-1}} \left(v_n - \gamma_C D^{-1}(v_n) + \gamma_C L \bar{x}_{n+1} \right). \end{cases} \tag{4.19}$$

(B) The fixed point formulation

We recall briefly the fixed point iteration corresponding to (4.19) whose detailed derivation can be found in [193, Section 3]. From the definition of the resolvent, it follows that the scheme (4.19) is equivalent to the following: for each $n \geq 0$,

$$\frac{1}{\gamma_A}(x_n - x_{n+1}) - B(x_n) - L^* v_n \in A(x_{n+1}),$$

$$\bar{x}_{n+1} = 2x_{n+1} - x_n,$$

$$\frac{1}{\gamma_C}(v_n - v_{n+1}) - D^{-1}(v_n) + L\bar{x}_{n+1} \in C^{-1}(v_{n+1}),$$

which can be written as follows:

$$-\begin{bmatrix} B & 0 \\ 0 & D^{-1} \end{bmatrix}\begin{pmatrix} x_n \\ v_n \end{pmatrix} \in \begin{bmatrix} A & L^* \\ -L & C^{-1} \end{bmatrix}\begin{pmatrix} x_{n+1} \\ v_{n+1} \end{pmatrix} + \begin{bmatrix} \mathrm{Id}_{\mathscr{H}}/\gamma_A & -L^* \\ -L & \mathrm{Id}_{\mathscr{G}}/\gamma_C \end{bmatrix}\begin{pmatrix} x_{n+1} - x_n \\ v_{n+1} - v_n \end{pmatrix},$$
(4.20)

where $\mathrm{Id}_{\mathscr{H}}$ and $\mathrm{Id}_{\mathscr{G}}$ are the identity operators on $\mathscr{H}$ and $\mathscr{G}$, respectively.

Define the product space $\mathscr{K} = \mathscr{H} \times \mathscr{G}$ and let Id be the identity operator on $\mathscr{K}$. Define the following variable and operators:

$$z_n := \begin{pmatrix} x_n \\ v_n \end{pmatrix}, \quad \mathbf{A} := \begin{bmatrix} A & L^* \\ -L & C^{-1} \end{bmatrix}, \quad \mathbf{B} := \begin{bmatrix} B & 0 \\ 0 & D^{-1} \end{bmatrix}, \quad \mathbf{V} := \begin{bmatrix} \mathrm{Id}_{\mathscr{H}}/\gamma_A & -L^* \\ -L & \mathrm{Id}_{\mathscr{G}}/\gamma_C \end{bmatrix}.$$
(4.21)

It follows that $\mathbf{A}$ is maximally monotone [28], $\mathbf{B}$ is η-cocoercive with $\eta = \min\{\beta_B, \beta_D\}$, and $\mathbf{V}$ is self-adjoint and ρ-positive definite with

$$\rho = (1 - \sqrt{\gamma_A \gamma_C \|L\|^2}) \min\left\{\frac{1}{\gamma_A}, \frac{1}{\gamma_C}\right\}$$

(see [48, 193]).

Let $\mathscr{K}_{\mathbf{V}}$ denote the Hilbert space induced by $\mathbf{V}$. Substituting the notions of (4.21) into (4.20), we get

$$-\mathbf{B}(z_n) \in \mathbf{A}(z_{n+1}) + \mathbf{V}(z_{n+1} - z_n),$$

which can be easily written as follows:

$$z_{n+1} = (\mathbf{V} + \mathbf{A})^{-1}(\mathbf{V} - \mathbf{B})(z_n) = (\mathbf{Id} + \mathbf{V}^{-1}\mathbf{A})^{-1}(\mathbf{Id} - \mathbf{V}^{-1}\mathbf{B})(z_n). \quad (4.22)$$

It can be seen that the scheme (4.22) is the forward-backward splitting method in $\mathscr{K}_{\mathbf{V}}$ (see [43, 48]). From the proof of [193, Theorem 3.1], it follows that $\mathbf{V}^{-1}\mathbf{A}$ is maximally monotone and $\mathbf{V}^{-1}\mathbf{B}$ is $\eta\rho$-cocoercive. By Lemma 4.1, the fixed point operator $\mathbf{T}_{\mathbf{FB}} = J_{\mathbf{V}^{-1}\mathbf{A}}(\mathbf{Id} - \mathbf{V}^{-1}\mathbf{B})$ is $\frac{2\eta\rho}{4\eta\rho - 1}$-averaged.

Chapter 5
The Inertial Krasnosel'skiĭ–Mann Iteration

The inertial type algorithms [172] originate from the heavy ball method of the so-called *heavy ball* with friction dynamical system:

$$\ddot{x}(t) + \gamma(t)\dot{x}(t) + \nabla\varphi(x(t)) = 0, \tag{5.1}$$

where φ is differentiable.

The system (5.1) arises when Newton's law is applied to a point subject to a constant friction $\gamma(t) > 0$ (of the velocity $\dot{x}(t)$) and a gravity potential φ.

An *explicit finite difference discretization* of the above system (5.1) can be constructed as follows: for each $n \geq 1$,

$$\frac{x_{n+1} - 2x_n + x_{n-1}}{h_n^2} + \gamma_n \frac{x_n - x_{n-1}}{h_n} + \nabla\varphi(x_n) = 0, \tag{5.2}$$

where $h_n > 0$ is time step and $\gamma_n = \gamma(t_n)$. From (5.2), it follows an iterative algorithm of the form:

$$x_{n+1} = x_n + \mu_n(x_n - x_{n-1}) - \nu_n \nabla\varphi(x_n) \quad \text{for each } n \geq 1,$$

where $\mu_n = 1 - \gamma_n h_n$ and $\nu_n = h_n^2$. The above iteration can be rewritten as follows: for each $n \geq 1$,

$$\begin{cases} z_n = x_n + \mu_n(x_n - x_{n-1}), \\ x_{n+1} = z_n - \nu_n \nabla\varphi(x_n). \end{cases} \tag{5.3}$$

Polyak [172] first called the scheme (5.3) as the *inertial type extrapolation algorithm*. The term $\mu_n(x_n - x_{n-1})$ is referred to as the *inertial term* and $\{\mu_n\}$ as the *inertial parameter sequence*.

© The Author(s), under exclusive license to Springer Nature Switzerland AG 2022
Q.-L. Dong et al., *The Krasnosel'skiĭ-Mann Iterative Method*, SpringerBriefs in Optimization, https://doi.org/10.1007/978-3-030-91654-1_5

The main feature of inertial type methods is that the next iterate is defined by making use of the previous two iterates. The inertial extrapolation technique has been widely used to accelerate the iterative algorithms from the fixed point problem, the convex optimization and nonconvex optimization since the cost of each iteration stays basically unchanged.

The popular accelerated gradient method of Nesterov [159] is given as follows: for each $n \geq 1$,

$$\begin{cases} z_n = x_n + \mu_n(x_n - x_{n-1}), \\ x_{n+1} = z_n - \nu_n \nabla \varphi(z_n). \end{cases} \tag{5.4}$$

The above scheme has some similarities with the heavy ball method (5.3), but it differs from the latter in one regard: while the heavy ball method uses gradients based on the current iterate, Nesterov's accelerated gradient method evaluates the gradient at points that are extrapolated by the inertial force.

The inertial parameter sequence $\{\mu_n\}$ is key to inertial type algorithms and a careful choice of the inertial parameter sequence is known to accelerate the theoretical functional convergence rate from $O(1/n)$ to $O(1/n^2)$ or $o(1/n^2)$ for a large class of algorithms (see, for example, [4, 17, 159]).

Recently, many authors have studied inertial algorithms for solving nonlinear problems, equilibrium problems, saddle point problems, variational inequality problems, split feasibility problems, optimization, and others in several ways (see [4, 6, 17, 34, 105, 106, 127, 159, 176] for details).

5.1　General Inertial Krasnosel'skiĭ–Mann Iterations

In this section, we consider the general inertial Krasnosel'skiĭ–Mann iterations and investigate the choices of the inertial parameters.

By using the techniques of the inertial extrapolation, Mainge [134] first introduced the *classical inertial Krasnosel'skiĭ–Mann iteration*: for each $n \geq 1$,

$$\begin{cases} y_n = x_n + \alpha_n(x_n - x_{n-1}), \\ x_{n+1} = (1 - \lambda_n)y_n + \lambda_n T(y_n), \end{cases} \tag{5.5}$$

and showed that the iterative sequence $\{x_n\}$ converges weakly to a fixed point of the nonexpansive mapping T provided that the inertial parameter sequence $\{\alpha_n\}$ and relaxation parameter sequence $\{\lambda_n\}$ satisfy:

(C1) $\{\alpha_n\} \subseteq [0, \alpha]$, where $\alpha \in [0, 1)$;
(C2) $\sum_{n=0}^{\infty} \alpha_n \|x_n - x_{n-1}\|^2 < +\infty$;
(C3) $\inf_{n\geq 0} \lambda_n > 0$ and $\sup_{n\geq 0} \lambda_n < 1$.

Since the summability condition (C2) involves the iterative sequence $\{x_n\}$, one needs to calculate the inertial parameter α_n per iteration (see more details in [155]).

To this end, Bot and Csetnek [24] got rid of the condition (C2) and replaced (C1) and (C3) with the following conditions, respectively:

(D1) $\{\alpha_n\} \subseteq [0, \alpha]$ is nondecreasing with $\alpha_1 = 0$ and $0 \leq \alpha < 1$;
(D2) For each $n \geq 1$,

$$\delta > \frac{\alpha^2(1+\alpha) + \alpha\sigma}{1 - \alpha^2}, \quad 0 < \lambda \leq \lambda_n \leq \frac{\delta - \alpha[\alpha(1+\alpha) + \alpha\delta + \sigma]}{\delta[1 + \alpha(1+\alpha) + \alpha\delta + \sigma]},$$

where $\lambda, \sigma, \delta > 0$.

Remark 5.1 We present two comments to the conditions (D1) and (D2) as follows:

(1) In (D2), the relaxation parameter sequence $\{\lambda_n\}$ is restricted by the constants δ, σ, and α. To show the impact of these constants to the relaxation parameters, Cui et al. [53] plotted the range of the relaxation parameters $\{\lambda_n\}$ in Fig. 5.1.
    Let $\delta = \frac{\alpha^2(1+\alpha)+\alpha\sigma}{1-\alpha^2} + \delta_*$. To get a large selection of the relaxation parameter sequence $\{\lambda_n\}$, they recommended to set $\sigma = 0.01$ or $\sigma = 0.001$ and $\delta_* = 1$. From Fig. 5.1, it is obvious that the range of the relaxation parameter sequence $\{\lambda_n\}$ are decreasing as the maximal inertial parameter α increasing.
(2) By putting together parameters given in Example 3.3 and the iteration (3.5), Combettes and Glaudin gave the same convergence conditions in [45, Example 4.4] as (D1) and (D2).

It was claimed in [74] that the condition (D2) is too complicated to determine upper bound of the inertial parameter sequence $\{\alpha_n\}$ in a simple way even if the relaxation parameter λ_n has been known. Moreover, in the case of $\lambda_n \equiv 0.5$, (D2) is restrictive. To circumvent such difficulty, a simple and smart way of choosing inertial parameters is proposed.

Let ε be any given sufficiently small positive number and, for each $n \geq 0$, set

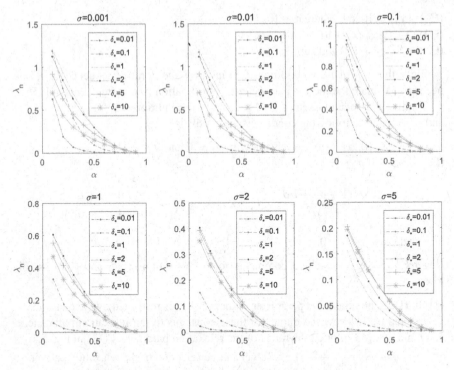

Fig. 5.1 The relaxation parameters λ_n versus the given constants of δ_*, σ and α

$$\Delta_n := \sqrt{\left(\frac{1}{\lambda_n} + \frac{1}{\lambda_{n+1}} - 1\right)^2 - 4\left(\frac{1}{\lambda_{n+1}} - 2\right)\left(\frac{1}{\lambda_n} - 1 - \epsilon\right)},$$

$$\rho(n) := \frac{1}{2}\frac{1}{\frac{1}{\lambda_{n+1}} - 2}\left(\frac{1}{\lambda_n} + \frac{1}{\lambda_{n+1}} - 1 - \Delta_n\right),$$

$$p_n := \frac{1}{2}\frac{1}{2 - \frac{1}{\lambda_{n+1}}}\left(\frac{1}{\lambda_n} + \frac{1}{\lambda_{n+1}} - 1\right),$$

$$q_n := \frac{1}{2 - \frac{1}{\lambda_{n+1}}}\left(\frac{1}{\lambda_n} - 1 - \epsilon\right),$$

$$\alpha(\lambda_n, \lambda_{n+1}, \epsilon) := \sqrt{p_n^2 + q_n} - p_n.$$

The following upper bound is presented in [74]:

$$\alpha_{n+1} \leq \rho(n), \quad \text{if } \lambda_n \in (0, 0.5) \text{ with } 1/\lambda_n - 1/\lambda_{n+1} + 3 > 0,$$

$$\alpha_{n+1} \leq (1 - \epsilon)/3, \quad \text{if } \lambda_n \equiv 0.5,$$

$$\alpha_{n+1} \leq \alpha(\lambda_n, \lambda_{n+1}, \epsilon), \quad \text{if } \lambda_n \in (0.5, 1 - \epsilon].$$

Very recently, Dong et al. [58] introduced the *general inertial Krasnosel'skiĭ–Mann iteration* as follows: for each $n \geq 1$,

$$\begin{cases} y_n = x_n + \alpha_n(x_n - x_{n-1}), \\ z_n = x_n + \beta_n(x_n - x_{n-1}), \\ x_{n+1} = (1 - \lambda_n)y_n + \lambda_n T(z_n), \end{cases} \quad (5.6)$$

where $\{\alpha_n\}, \{\beta_n\}, \{\lambda_n\} \subseteq [0, 1]$.

Remark 5.2 By its form, the general inertial Krasnosel'skiĭ–Mann iteration is the most general Krasnosel'skiĭ–Mann iteration with inertial effects. It is easy to show that the general inertial Krasnosel'skiĭ–Mann iteration includes other inertial Krasnosel'skiĭ–Mann iterations as special cases.

The relations between the iteration (5.6) with other work are as follows:

(1) $\alpha_n = \beta_n$ for each $n \geq 1$, i.e., $y_n = z_n$: this is the classical *inertial Krasnosel'skiĭ–Mann iteration* (5.5);
(2) $\beta_n = 0$ for each $n \geq 1$: this becomes the *accelerated Krasnosel'skiĭ–Mann iteration* introduced in [72]: for each $n \geq 1$,

$$\begin{cases} y_n = x_n + \alpha_n(x_n - x_{n-1}), \\ x_{n+1} = \lambda_n y_n + (1 - \lambda_n)T(x_n). \end{cases} \quad (5.7)$$

The scheme (5.7) can be seen the explicit discretization of the following dynamical system:

$$\ddot{x}(t) + \gamma(t)\dot{x}(t) + (\text{Id} - T)(x(t)) = 0.$$

Indeed, similar to (5.3), we obtain

$$x_{n+1} = x_n + \mu_n(x_n - x_{n-1}) + \nu_n(T(x_n) - x_n) \quad \text{for each } n \geq 1. \quad (5.8)$$

Let $\lambda_n = 1 - \nu_n$ and $\alpha_n = \frac{\mu_n}{1 - \nu_n}$. Then (5.8) becomes (5.7).

(3) $\alpha_n = 0$ for each $n \geq 1$: it becomes the following algorithm: for each $n \geq 1$,

$$\begin{cases} z_n = x_n + \beta_n(x_n - x_{n-1}), \\ x_{n+1} = (1 - \lambda_n)x_n + \lambda_n T(z_n). \end{cases} \quad (5.9)$$

Inspired by Malitsky [138] and Mainge [135], the algorithm (5.9) is called the *reflected Krasnosel'skiĭ–Mann iteration*.

For *L*-Lipschitz pseudocontractive mappings, Moudafi [153] introduced the following *reflected inertial Krasnosel'skiĭ–Mann iteration*:

$$x_{n+1} = \lambda x_{n-1} + (1 - 2\lambda)x_n + \lambda T(2x_n - x_{n-1}) \quad \text{for each } n \geq 1.$$

The above scheme is obviously a special case of the general inertial Krasnosel'skiĭ–Mann iteration (5.6) with the following conditions:

$$\lambda_n \equiv \lambda \in \left(0, \frac{\sqrt{2}-1}{L+1}\right), \quad \alpha_n = \frac{\lambda}{1-\lambda}, \quad \beta_n \equiv 1 \quad \text{for each } n \geq 1.$$

The convergence conditions of the general inertial Krasnosel'skiĭ–Mann iteration (5.6) are given as follows:

(E1) The sequences $\{\alpha_n\} \subset [0, \alpha]$ and $\{\beta_n\} \subset [0, \beta]$ are nondecreasing with $\alpha_1 = \beta_1 = 0, \alpha, \beta \in [0, 1)$ and $\lambda_n \equiv \lambda$;
(E2) For any $\sigma, \delta > 0$,

$$\delta > \frac{\alpha\xi(1+\xi) + \alpha\sigma}{1-\alpha^2}, \quad 0 < \lambda \leq \frac{\delta - \alpha[\xi(1+\xi) + \alpha\delta + \sigma]}{\delta[1 + \xi(1+\xi) + \alpha\delta + \sigma]}, \quad (5.10)$$

where $\xi = \max\{\alpha, \beta\}$.

Theorem 5.1 ([58, Theorem 1]) *Let* $T : \mathscr{H} \to \mathscr{H}$ *be a nonexpansive mapping with* $Fix(T) \neq \emptyset$. *Assume that the conditions* (E1) *and* (E2) *hold. Then the sequence* $\{x_n\}$ *generated by the general inertial Krasnosel'skiĭ–Mann iteration* (5.6) *converges weakly to a point in* $Fix(T)$.

Let $\beta_n = 0$ for each $n \geq 1$. Then we obtain the convergence condition of the accelerated Krasnosel'skiĭ–Mann iteration:

Theorem 5.2 *Let* $T : \mathscr{H} \to \mathscr{H}$ *be a nonexpansive mapping with* $Fix(T) \neq \emptyset$. *Assume that the sequence* $\{\alpha_n\} \subset [0, \alpha]$ *is nondecreasing with* $\alpha_1 = 0, 0 \leq \alpha < 1$ *and* $\lambda_n \equiv \lambda$ *satisfies*

$$\delta > \frac{\alpha^2(1+\alpha) + \alpha\sigma}{1-\alpha^2}, \quad 0 < \lambda \leq \frac{\delta - \alpha[\alpha(1+\alpha) + \alpha\delta + \sigma]}{\delta[1 + \alpha(1+\alpha) + \alpha\delta + \sigma]},$$

where $\sigma, \delta > 0$. *Then the sequence* $\{x_n\}$ *generated by the accelerated Krasnosel'skiĭ–Mann iteration* (5.7) *converges weakly to a point in* $Fix(T)$.

Let $\alpha_n = 0$ for each $n \geq 1$. Then we obtain the convergence theorem of the reflected Krasnosel'skiĭ–Mann iteration:

Theorem 5.3 *Let* $T : \mathcal{H} \to \mathcal{H}$ *be a nonexpansive mapping with* $FixT \neq \emptyset$. *Assume that the sequence* $\{\beta_n\} \subset [0, \beta]$ *is nondecreasing with* $\beta_1 = 0, 0 \leq \beta < 1$ *and* $\lambda_n \equiv \lambda$ *satisfies*

$$0 < \lambda \leq \frac{1}{1 + \beta(1 + \beta) + \sigma},\tag{5.11}$$

where $\sigma > 0$. *Then the sequence* $\{x_n\}$ *generated by the reflected Krasnosel'skiĭ– Mann iteration* (5.9) *converges weakly to a point in* $Fix(T)$.

As stated in [74], the condition (E2) is complicated since it involves constants σ and δ and, furthermore, the condition (E2) is restrictive.

Now, we relax the upper bound in (5.11) by using the technique in [74]. First, we introduce a lemma which is key to the main result:

Lemma 5.1 *Let* $T : \mathcal{H} \to \mathcal{H}$ *be a nonexpansive mapping with* $Fix(T) \neq \emptyset$ *and* $\{x_n\}$ *be the sequence generated by the reflected inertial Krasnosel'skiĭ–Mann iteration* (5.9). *Then, for any given fixed point* p *of* T, *we have*

$$\|x_{n+1} - p\|^2 - (1 + \theta_n)\|x_n - p\|^2 + \theta_n\|x_{n-1} - p\|^2$$
$$\leq \left(1 - \frac{1}{\lambda_n}\right)\|x_{n+1} - x_n\|^2 + \mu_n\|x_n - x_{n-1}\|^2,\tag{5.12}$$

where

$$\theta_n = \beta_n\lambda_n, \quad \mu_n = \lambda_n\beta_n(1 + \beta_n) \text{ for each } n \geq 1.\tag{5.13}$$

Proof From (2.4), it follows that

$$\|x_{n+1} - p\|^2 = (1 - \lambda_n)\|x_n - p\|^2 + \lambda_n\|Tz_n - p\|^2$$
$$- \lambda_n(1 - \lambda_n)\|Tz_n - x_n\|^2$$
$$\leq (1 - \lambda_n)\|x_n - p\|^2 + \lambda_n\|z_n - p\|^2 - \lambda_n(1 - \lambda_n)\|Tz_n - x_n\|^2.\tag{5.14}$$

Using (2.4) again, we have

$$\|z_n - p\|^2 = \|(1 + \beta_n)(x_n - p) - \beta_n(x_{n-1} - p)\|^2$$
$$= (1 + \beta_n)\|x_n - p\|^2 - \beta_n\|x_{n-1} - p\|^2 + \beta_n(1 + \beta_n)\|x_n - x_{n-1}\|^2.\tag{5.15}$$

Combining (5.14) and (5.15), we have

$$\|x_{n+1} - p\|^2 - (1 + \theta_n)\|x_n - p\|^2 + \theta_n\|x_{n-1} - p\|^2$$
$$\leq -\lambda_n(1 - \lambda_n)\|Tz_n - x_n\|^2 + \lambda_n\beta_n(1 + \beta_n)\|x_n - x_{n-1}\|^2,\tag{5.16}$$

where θ_n is defined in (5.13). Using (5.9), we have

$$\|T z_n - x_n\| = \frac{1}{\lambda_n}\|x_{n+1} - x_n\|. \tag{5.17}$$

Therefore, from (5.16) and (5.17), we can derive the inequality (5.12). This completes the proof.

Theorem 5.4 *Let* $T : \mathcal{H} \to \mathcal{H}$ *be a nonexpansive mapping with* $\operatorname{Fix} T \neq \emptyset$. *Assume that the sequences* $\{\lambda_n\} \subset [0, 1]$ *and* $\{\beta_n\} \subset [0, \beta]$ *is nondecreasing with* $\beta_1 = 0, 0 < \beta < 1$ *and satisfy:*

$$0 < \lambda \le \lambda_n \le \frac{2}{1 + \sqrt{1 + 4\beta(1 + \beta)} + \sigma}, \tag{5.18}$$

where $\lambda, \sigma > 0$. *Then the sequence* $\{x_n\}$ *generated by the reflected Krasnosel'skiĭ–Mann iteration* (5.9) *converges weakly to a point in* $\operatorname{Fix}(T)$.

Proof Note that the sequence $\{\theta_n\}$ is nondecreasing since $\{\beta_n\}$ and $\{\lambda_n\}$ are nondecreasing. Define the sequences $\{\phi_n\}$ and $\{\Psi_n\}$ by

$$\phi_n := \|x_n - p\|^2 \quad \text{for each } n \ge 1$$

and

$$\Psi_n := \phi_n - \theta_n \phi_{n-1} + \mu_n \|x_n - x_{n-1}\|^2 \quad \text{for each } n \ge 1,$$

respectively. Using the monotonicity of $\{\theta_n\}$ and the fact that $\phi_n \ge 0$ for each $n \ge 1$, we have

$$\Psi_{n+1} - \Psi_n \le \phi_{n+1} - (1+\theta_n)\phi_n + \theta_n\phi_{n-1} + \mu_{n+1}\|x_{n+1} - x_n\|^2 - \mu_n\|x_n - x_{n-1}\|^2.$$

By (5.12), we know

$$\Psi_{n+1} - \Psi_n \le \left(1 - \frac{1}{\lambda_n} + \mu_{n+1}\right)\|x_{n+1} - x_n\|^2$$
$$\le \left(1 - \frac{1}{\lambda_{n+1}} + \mu_{n+1}\right)\|x_{n+1} - x_n\|^2, \tag{5.19}$$

where the second inequality comes from the fact that $\{\lambda_n\}$ is nondecreasing. From (5.18), it is easy to verify that

$$-\frac{1 - \lambda_n}{\lambda_n} + \lambda_n\beta_n(1 + \beta_n) \le -\sigma \quad \text{for each } n \ge 1. \tag{5.20}$$

It follows from (5.19) and (5.20) that

$$\Psi_{n+1} - \Psi_n \leq -\sigma \|x_{n+1} - x_n\|^2 \quad \text{for each } n \geq 1. \tag{5.21}$$

The rest of the proof is similar to that of Theorem 1 in [58] and omitted here. This completes the proof.

The range of the relaxation parameter sequence $\{\lambda_n\}$ in (5.11) is larger than that in (5.18) for the sufficiently small positive number σ since

$$1 + 2\beta(1+\beta) > \sqrt{1 + 4\beta(1+\beta)} \quad \text{for all } \beta \in (0, 1].$$

Note that the inertial parameters α_n and β_n satisfying (E1) are nonnegative. Dong et al. [66] relaxed the ranges of α_n and β_n and gave further results, for which α_n and β_n can be taken negative. They considered two cases as follows:

Case 1: $\alpha_n \in [0, 1]$ and $\beta_n \in (-\infty, 0]$;
Case 2: $\alpha_n \in [-1, 0]$ and $\beta_n \in [0, +\infty)$.

The convergence results are given in following theorems:

Theorem 5.5 *Let $T : \mathcal{H} \to \mathcal{H}$ be a nonexpansive mapping with $\mathrm{Fix}(T) \neq \emptyset$. Assume that the sequences $\{\alpha_n\}$, $\{\beta_n\}$, and $\{\lambda_n\}$ satisfy the following conditions:*

(E3) $\alpha_1 = \beta_1 = 0$, $\{\alpha_n\} \subset [\underline{\alpha}, \overline{\alpha}]$ *and* $\{\beta_n\} \subset [\underline{\beta}, \overline{\beta}]$ *are nondecreasing and*
$\quad \underline{\alpha}, \overline{\alpha} \in [0, 1)$, $\underline{\beta}, \overline{\beta} \in (-\infty, -1]$, $\lambda_n \equiv \lambda$;
(E4) *For any $\sigma, \delta > 0$,*

$$\delta > \frac{\overline{\alpha}(\xi + \sigma)}{1 - \overline{\alpha}^2},$$

where

$$\xi = \max\{\overline{\alpha}(1+\overline{\alpha}), \underline{\beta}(1+\underline{\beta})\}$$

and

$$0 < \lambda \leq \min\left\{ \frac{\underline{\alpha}}{\underline{\alpha} - \underline{\beta}}, \frac{\delta - \overline{\alpha}(\xi + \overline{\alpha}\delta + \sigma)}{\delta(\xi + \overline{\alpha}\delta + \sigma + 1)} \right\}.$$

Then the sequence $\{x_n\}$ generated by the general inertial Krasnosel'skiĭ–Mann iteration (5.6) converges weakly to a point in $\mathrm{Fix}(T)$.

Theorem 5.6 *Let $T : \mathcal{H} \to \mathcal{H}$ be a nonexpansive mapping with $\mathrm{Fix}(T) \neq \emptyset$. Assume that the sequences $\{\alpha_n\}$, $\{\beta_n\}$, and $\{\lambda_n\}$ satisfy the following conditions:*

(E5) $\alpha_1 = \beta_1 = 0$, $\{\alpha_n\} \subset [\underline{\alpha}, \overline{\alpha}]$ *and* $\{\beta_n\} \subset [\underline{\beta}, \overline{\beta}]$ *are nondecreasing and*
$\quad \underline{\alpha}, \overline{\alpha} \in (0, 1)$, $\underline{\beta}, \overline{\beta} \in [-1, 0)$, $\lambda_n \equiv \lambda$;
(E6) *For any $\sigma, \delta > 0$,*

$$\delta > \frac{\overline{\alpha}[\overline{\alpha}(1+\overline{\alpha})+\sigma]}{1-\overline{\alpha}^2},$$

where

$$0 < \lambda \le \min\left\{\frac{\alpha}{\underline{\alpha}-\underline{\beta}}, \frac{\underline{\alpha}(1+\underline{\alpha}+\delta)}{\underline{\alpha}(1+\underline{\alpha}+\delta)-\eta}, \frac{\delta-\overline{\alpha}[\overline{\alpha}(1+\overline{\alpha}+\delta)+\sigma]}{\delta[\overline{\alpha}(1+\overline{\alpha}+\delta)+\sigma+1]}\right\}$$

and

$$\eta = \min_{\beta\in[\underline{\beta},\overline{\beta}]}\{\beta(1+\beta)\}.$$

Then the sequence $\{x_n\}$ generated by the general inertial Krasnosel'skiĭ–Mann iteration (5.6) converges weakly to a point in Fix(T).

Theorem 5.7 *Let $T : \mathscr{H} \to \mathscr{H}$ be a nonexpansive mapping with Fix(T) ≠ ∅. Assume that the sequences $\{\alpha_n\}$, $\{\beta_n\}$, and $\{\lambda_n\}$ satisfy the following conditions:*

(E7) *$\alpha_1 = \beta_1 = 0$ and $\{\alpha_n\} \subset [\underline{\alpha}, \overline{\alpha}]$ and $\{\beta_n\} \subset [\underline{\beta}, \overline{\beta}]$ are nondecreasing and $\underline{\alpha}, \overline{\alpha} \in (-1, 0]$, $\underline{\beta}, \overline{\beta} \in (0, +\infty)$, $\lambda_n \equiv \lambda$;*
(E8) *For any $\sigma, \delta > 0$,*

$$\delta > \max\left\{1+\overline{\alpha}, \frac{\underline{\alpha}[\overline{\beta}(1+\overline{\beta})+\sigma]}{\underline{\alpha}^2-1}\right\},$$

where

$$\frac{\underline{\alpha}}{\underline{\alpha}-\underline{\beta}} \le \lambda \le \min\left\{\frac{1-\overline{\alpha}}{\overline{\beta}-\overline{\alpha}}, \frac{\delta+\underline{\alpha}[\overline{\beta}(1+\overline{\beta})-\underline{\alpha}\delta+\sigma]}{\delta[\overline{\beta}(1+\overline{\beta})-\underline{\alpha}\delta+\sigma+1]}\right\}.$$

Then the sequence $\{x_n\}$ generated by the general inertial Krasnosel'skiĭ–Mann iteration (5.6) converges weakly to a point in Fix(T).

By using the relation of the Krasnosel'skiĭ–Mann iteration (5.6) and operators splitting methods, it is easy to extend and generalize the results on the inertial Krasnosel'skiĭ–Mann iteration (5.6) to the inertial operators splitting methods.

Bot et al. [23, 24] introduced the inertial Douglas–Rachford splitting and the inertial forward-backward-forward primal-dual splitting algorithm. Recently, Cui et al. [53] introduced the *inertial Davis–Yin splitting* as follows: for each $n \ge 1$,

$$\begin{cases} w_n = x_n + \alpha_n(x_n - x_{n-1}), \\ y_n = J_{\gamma A_2} w_n, \\ z_n = J_{\gamma A_1}(2y_n - w_n - \gamma B y_n), \\ x_{n+1} = w_n + \lambda_n(z_n - y_n), \end{cases} \tag{5.22}$$

where $A_1, A_2 : \mathcal{H} \rightrightarrows \mathcal{H}$ are maximally monotone and $B : \mathcal{H} \to \mathcal{H}$ is β-cocoercive for some $\beta > 0$ and $\gamma \in (0, 2\beta)$. The iteration (5.22) equals to the following: for each $n \geq 1$,

$$\begin{cases} w_n = x_n + \alpha_n(x_n - x_{n-1}), \\ x_{n+1} = (1 - \lambda_n)w_n + \lambda_n T_{DY}(w_n), \end{cases} \tag{5.23}$$

where T_{DY} is defined by (4.13).

The iteration (5.23) can be seen as a special case of the classical inertial Krasnosel'skiĭ–Mann iteration (5.5). The convergence result of the inertial Davis–Yin splitting is presented in the following theorem:

Theorem 5.8 ([53, Theorem 3.1]) *Let* $A_1, A_2 : \mathcal{H} \rightrightarrows \mathcal{H}$ *be two maximally monotone operators and* $B : \mathcal{H} \to \mathcal{H}$ *be a* β-*cocoercive mapping for some* $\beta > 0$. *Assume that the parameters* γ, $\{\alpha_n\}$ *and* $\{\lambda_n\}$ *satisfy the following conditions:*

(F1) $\gamma \in (0, 2\beta\epsilon)$, *where* $\epsilon \in (0, 1)$;
(F2) $\{\alpha_n\}$ *is nondecreasing with* $\alpha_1 = 0$ *and* $0 \leq \alpha_n \leq \alpha < 1$;
(F3) *For each* $n \geq 0$ *and* λ, σ, $\delta > 0$,

$$\delta > \frac{\alpha^2(1 + \alpha) + \alpha\sigma}{1 - \alpha^2}, \quad 0 < \lambda \leq \lambda_n \leq \frac{\delta - \alpha[\alpha(1 + \alpha) + \alpha\delta + \sigma]}{\overline{\alpha}\delta[1 + \alpha(1 + \alpha) + \alpha\delta + \sigma]},$$

where $\overline{\alpha} = \frac{1}{2 - \epsilon}$.

Then the sequence $\{x_n\}$ *generated by the inertial Davis–Yin splitting (5.22) converges weakly to a fixed point of* T_{DY}.

Note that Cevher [33] independently introduced the inertial Davis–Yin splitting (5.22) and presented different conditions on the inertial parameter $\{\alpha_n\}$ and relaxation parameter $\{\lambda_n\}$.

5.2   The Alternated Inertial Krasnosel'skiĭ–Mann Iteration and the Online Inertial Krasnosel'skiĭ–Mann Iteration

The inertial term generally loses the monotonicity of the sequence of the distance between the iterative sequence and a solution, which may lead to the slow convergence of the inertial algorithms (see, for example, [140]). Recently, the alternated inertial methods which apply inertia every other iteration was introduced

[157], whose error monotonously decreases again. The alternated inertial algorithms show good convergence properties and numerical performances (see, for example, [115, 180, 181]), along with its remarkable closeness with relaxation.

Iutzeler and Hendrickx [114] introduced the alternated inertial Picard iteration for an averaged mapping, which can be seen as the following alternated inertial Krasnosel'skiĭ–Mann iteration with the fixed relaxation parameter λ and the nonexpansive mapping T:

$$x_{n+1} = (1 - \lambda)y_n + \lambda T(y_n) \quad \text{for each } n \geq 1, \tag{5.24}$$

where $\lambda \in (0, 1)$ and

$$y_n = \begin{cases} x_n, & \text{if } n = \text{even}, \\ x_n + \alpha_n(x_n - x_{n-1}), & \text{if } n = \text{odd}. \end{cases} \tag{5.25}$$

The convergence result of the alternated inertial Krasnosel'skiĭ–Mann iteration (5.24)-(5.25) is given as follows:

Theorem 5.9 ([114, Lemma 3.3]) *Let* $T : \mathbb{R}^N \to \mathbb{R}^N$ *be a nonexpansive operator such that* $Fix(T) \neq \emptyset$. *Assume that* $\lambda \in (0, 1)$ *and the sequence* $\{\alpha_n\}$ *verify*

$$0 \leq \alpha_n \leq \frac{1 - \lambda}{\lambda} \quad \text{for each } n \geq 1.$$

Then the sequence $\{x_n\}$ *generated by the alternated inertial Krasnosel'skiĭ–Mann iteration (5.24)-(5.25) converges to a point in* $Fix(T)$.

Iutzeler and Hendrickx [114] also introduced the online inertial method and the online alternated inertial method, which automatically tune the inertial parameters online.

A restart mechanism has to be introduced to make sure that the algorithm converges since this scheme often overpasses the theoretical limits of the convergence results. Thus, in order to maintain convergence, the algorithm must either

(1) sufficiently decrease the error $x_n - y_n$
   or
(2) set the inertial parameter α_n to 0 so that classical convergence results apply.

Note that the online inertial Krasnosel'skiĭ–Mann iteration uses the same inertial parameter α_{2n} in the iterates x_{2n+1} and x_{2n+2}, which is given adaptively.

The next theorem gives the convergence of the online inertial Krasnosel'skiĭ–Mann iteration:

Theorem 5.10 ([114, Theorem 4.2]) *Let* $T : \mathbb{R}^N \to \mathbb{R}^N$ *be a nonexpansive operator such that* $Fix(T) \neq \emptyset$. *Then the sequence* $\{y_n\}$ *generated by the online inertial Krasnosel'skiĭ–Mann iteration converges in the sense that*

Algorithm 1 The online inertial Krasnosel'skiĭ–Mann iteration

---

Initialization: x_1, $x_2 = (1 - \lambda)x_1 + \lambda T(x_1)$, $y_2 = x_1$, $\alpha_2 = 0$, $\varepsilon > 0$.
For each $n \geq 1$, set

$$
\begin{cases}
y_{2n+1} = x_{2n} + \alpha_{2n}(x_{2n} - x_{2n-1}), \\
x_{2n+1} = (1 - \lambda)y_{2n+1} + \lambda T(y_{2n+1}), \\
y_{2n+2} = x_{2n+1} + \alpha_{2n}(x_{2n+1} - x_{2n}), \\
x_{2n+2} = (1 - \lambda)y_{2n+2} + \lambda T(y_{2n+2}).
\end{cases}
$$

Let $c_{2n} = \max \left\{ \dfrac{\|x_{2n+2} - y_{2n+2}\|}{\|x_{2n+1} - y_{2n+1}\|}, \dfrac{\|x_{2n+1} - y_{2n+1}\|}{\|x_{2n} - y_{2n}\|} \right\}$ for each $n \geq 1$.

if $c_{2n} \leq 1 - \varepsilon$ [Acceleration]

$$
v_{2n} = \sqrt{\frac{\|x_{2n+2} - x_{2n+1}\|^2 + \|x_{2n+1} - x_{2n}\|^2}{\|x_{2n+1} - x_{2n}\|^2 + \|x_{2n} - x_{2n-1}\|^2}},
$$

$$
\theta_{2n} = \min \left\{ \frac{(v_{2n})^2}{\alpha_{2n}v_{2n} - \alpha_{2n} + v_{2n}}, 1 - \varepsilon \right\},
$$

$$
\alpha_{2n+2} = \max \left\{ 0, \frac{(1 - \sqrt{1 - \theta_{2n}})^2}{\theta_{2n}} \right\}.
$$

elseif $\alpha_{2n} > 0$ [Restart]

$$
\alpha_{2n+2} = 0,
$$

$$
(x_{2n+1}, x_{2n+2}, y_{2n+2}) = (x_{2n-1}, x_{2n}, y_{2n}).
$$

elseif $\alpha_{2n} = 0$ [No Acceleration]

$$
\alpha_{2n+2} = 0.
$$

$$
\|T(y_n) - y_n\| \to 0.
$$

Furthermore, if Fix(T) is reduced to a single point x^*, then $x_n \to x^*$.

When the convergence is sublinear, the restart condition based on a constant may be too harsh. One can deduce that ϵ can be taken as a sequence (ϵ^ℓ) provided that $1/\epsilon^\ell = o(\ell)$ where ℓ is the number of accelerations. For instance, a typical setting is to keep track of the number of accelerations ℓ and take $\epsilon^\ell = \epsilon_0/\sqrt{\ell}$. Note that in the sublinear case, the online inertial Krasnosel'skiĭ–Mann iteration makes the acceleration parameter go to 1 as in Nesterov's optimal method [159].

Similar to the online inertial Krasnosel'skiĭ–Mann iteration, the online alternated inertial Krasnosel'skiĭ–Mann iteration uses the same inertial parameter α_{4n} in the iterates x_{4n+1} and x_{4n+3}, which is given adaptively.

Algorithm 2 The online alternated inertial Krasnosel'skiĭ–Mann iteration

---

Initialization: $x_3, x_4 = (1 - \lambda)x_3 + \lambda T(x_3), y_4 = x_3, \alpha_4 = 0, \varepsilon > 0.$
For each $n \geq 1$: set

$$
\begin{cases}
y_{4n+1} = x_{4n} + \alpha_{4n}(x_{4n} - x_{4n-1}), \\
x_{4n+1} = (1 - \lambda)y_{4n+1} + \lambda T(y_{4n+1}), \\
y_{4n+2} = x_{4n+1}, \\
x_{4n+2} = (1 - \lambda)y_{4n+2} + \lambda T(y_{4n+2}),
\end{cases}
\quad
\begin{cases}
y_{4n+3} = x_{4n+2} + \alpha_{4n}(x_{4n+2} - x_{4n+1}), \\
x_{4n+3} = (1 - \lambda)y_{4n+3} + \lambda T(y_{4n+3}), \\
y_{4n+4} = x_{4n+3}, \\
x_{4n+4} = (1 - \lambda)y_{4n+4} + \lambda T(y_{4n+4}).
\end{cases}
$$

Let $c_{4n} = \max \left\{ \dfrac{\|x_{4n+4} - x_{4n+3}\|}{\|x_{4n+2} - x_{4n+1}\|}, \dfrac{\|x_{4n+2} - x_{4n+1}\|}{\|x_{4n} - x_{4n-1}\|} \right\}$ for each $n \geq 1.$

if $c_{4n} \leq 1 - \varepsilon$ [Acceleration]

$$
v_{4n} = \frac{\|x_{4n+4} - x_{4n+2}\|^2}{\|x_{4n+2} - x_{4n}\|^2},
$$

$$
\theta_{4n} = \min \left\{ \frac{\alpha_{4n} + \sqrt{(\alpha_{4n})^2 + 4\alpha_{4n}v_{4n} + 4v_{4n}}}{2(\alpha_{4n} + 1)}, 1 - \varepsilon \right\},
$$

$$
\alpha_{4n+4} = \frac{2(\theta_{4n})^2 + (\sqrt{2} - 1)\theta_{4n}}{2\theta_{4n}(1 - \theta_{4n}) + \frac{1}{2}}.
$$

elseif $\alpha_{4n} > 0$ [Restart]

$$
\alpha_{4n+4} = 0,
$$

$$
(x_{4n+3}, x_{4n+4}) = (x_{4n-1}, x_{4n}).
$$

elseif $\alpha_{4n} = 0$ [No Acceleration]

$$
\alpha_{4n+4} = 0.
$$

Theorem 5.11 ([114, Theorem 4.3]) *Let* $T : \mathbb{R}^N \to \mathbb{R}^N$ *be a nonexpansive operator such that* $Fix(T) \neq \emptyset$. *Then the sequence* $\{x_n\}$ *generated by the online alternated inertial Krasnosel'skiĭ–Mann iteration converges in the sense that* $\|T(x_{2n}) - x_{2n}\| \to 0$. *Furthermore, if* $Fix(T)$ *is reduced to a single point* x^*, *then* $x_n \to x^*$.

Although various selection methods and several ranges of the inertial parameters have been discussed, there is few progress in the best choices. Therefore, herein we restate the famous open problem raised by Moudafi [155] and used it to the inertial Krasnosel'skiĭ–Mann iterations:

Open Problem *Which are the best choices of the inertial parameters of the inertial Krasnosel'skiĭ–Mann iterations from the view of the theoretical as well as numerical acceleration of the convergence?*

Anderson acceleration named after its inventor [2] was first proposed to speed up the iterative methods and has been widely used to accelerate the self-consistent field iteration in electronic structure computation. Recently, some researchers accelerated first order methods, such as the Douglas–Rachford splitting method and the alternating direction method of multipliers by using Anderson acceleration (see, for example, [1, 83, 165]). It is easy to show that Anderson acceleration is actually an inertial type method as its depth is one. Therefore, one interesting topic is to investigate the optimal choices of inertial parameters of the inertial Krasnosel'skiĭ–Mann iteration via Anderson acceleration.

Chapter 6
The Multi-step Inertial Krasnosel'skiĭ–Mann Iteration

In the book [164], Ortega and Rheinboldt introduced a general iterative process:

$$x_{n+1} = \Theta_n(x_n, x_{n-1}, \cdots, x_{n-s+1}) \quad \text{for each } n \geq 1, \tag{6.1}$$

where $s \geq 1$ is an integer and $\Theta_n(\cdot)$ is the function that performs "extrapolation" onto the points $x_n, x_{n-1}, \cdots, x_{n-s+1}$. The iterative process (6.1) is called s-*step method*.

Polyak [173] also suggested the idea of using more than 2 steps, i.e., the iterative process (6.1) with $s > 2$, to improve the convergence. However, neither the convergence nor the rate result is provided in [173].

In his Ph.D. dissertation [125], Liang firstly introduced the variable metric multi-step inertial forward–backward splitting method with an enlargement error. Let the variable metric be the identity operator and the enlargement error be zero. Then Liang's method can be addressed as follows:

$$\begin{cases} y_n = x_n + \sum_{k \in S} a_{n,k}(x_{n-k} - x_{n-k-1}), \\[2mm] z_n = x_n + \sum_{k \in S} b_{n,k}(x_{n-k} - x_{n-k-1}), \\[2mm] x_{n+1} = J_{\gamma A}(y_n - \gamma B z_n), \end{cases} \tag{6.2}$$

where $S = \{0, \cdots, s - 1\}$, $s \geq 1$, $A : \mathscr{H} \rightrightarrows \mathscr{H}$ is maximally monotone, and $B : \mathscr{H} \to \mathscr{H}$ is β-cocoercive for some $\beta > 0$ and $\gamma \in (0, 2\beta)$.

Set $\zeta_{n,k} := a_{n,k} - \frac{\gamma b_{n,k}}{2\beta}$ and $\bar{a}_k := \sup_{n \in \mathbb{N}} |a_{n,k}|$, $k \in S$. The iterative sequence $\{x_n\}$ generated by (6.2) weakly converges to a point in $\mathrm{zer}(A + B)$, provided that $\sum_{n \in S} \bar{a}_n < 1$ and

$$\sum_{n=0}^{+\infty} \max \left\{ \max_{k\in S} \zeta_{n,k}^2, \max_{k\in S} |a_{n,k}|, \max_{k\in S} |b_{n,k}| \right\} \sum_{k\in S} \|x_{n-k} - x_{n-k-1}\|^2 < +\infty.$$

Liang also presented the multi-step inertial primal–dual splitting method and the multi-step inertial Douglas–Rachford splitting method. Similar to the s-step method (6.1), the $(n + 1)$-th iterate x_{n+1} in Liang's method involves at most $s + 1$ previous iterations $x_n, x_{n-1}, \cdots, x_{n-s}$.

Inspired by the work in [125], Zhang et al. [204] recently proposed the following *multi-step inertial proximal contraction algorithm*:

Let $A : \mathcal{H} \rightrightarrows \mathcal{H}$ be maximally monotone and $B : \mathcal{H} \to \mathcal{H}$ be monotone and L-Lipschitz continuous. Let $s \in \mathbb{N}_+$ and $S := \{0, \cdots, s - 1\}$. Let $\{a_{n,k}\}_{n\in\mathbb{N}, k\in S} \in (-1, 1)^s$. Choose $x_0 \in \mathcal{H}$, $x_{-i-1} = x_0$, $i \in S\backslash\{0\}$ and, for each $n \geq 1$, apply the following iteration:

$$\begin{cases} \omega_n = x_n + \displaystyle\sum_{k\in S} a_{n,k}(x_{n-k} - x_{n-k-1}), \\[2mm] y_n = J_{\lambda_n A}(\omega_n - \lambda_n B(\omega_n)), \\[2mm] d(\omega_n, y_n) = (\omega_n - y_n) - \lambda_n(B(\omega_n) - B(y_n)), \\[2mm] x_{n+1} = \omega_n - \gamma\beta_n d(\omega_n, y_n), \end{cases} \qquad (6.3)$$

where $\gamma \in (0, 2)$ and

$$\beta_n := \begin{cases} \dfrac{\phi(\omega_n, y_n)}{\|d(\omega_n, y_n)\|^2}, & \text{if } \|d(\omega_n, y_n)\| \neq 0, \\[3mm] c, & \text{if } \|d(\omega_n, y_n)\| = 0, \end{cases}$$

where $c > 1$ is an arbitrary constant and $\phi(\omega_n, y_n) := \langle \omega_n - y_n, d(\omega_n, y_n) \rangle$.

The convergence of the algorithm (6.3) is given as follows:

Theorem 6.1 *Let $A : \mathcal{H} \rightrightarrows \mathcal{H}$ be maximally monotone and $B : \mathcal{H} \to \mathcal{H}$ be monotone and L-Lipschitz continuous on $\mathcal{H}$. Assume that $zer(A+B) \neq \emptyset$. Let $\lambda_n \in (0, \frac{1}{L})$. Suppose that $\sum_{k\in S} \overline{a}_k < 1$, where $\overline{a}_k := \sup_{n\in\mathbb{N}} |a_{n,k}|$. Then the sequence $\{x_n\}$ generated by the multi-step inertial proximal contraction algorithm (6.3) is bounded. If, moreover, the following summability condition holds:*

$$\sum_{n=0}^{+\infty} \max_{k\in S} |a_{n,k}| \sum_{k\in S} \|x_{n-k} - x_{n-k-1}\|^2 < +\infty, \qquad (6.4)$$

then the sequences $\{x_n\}$, $\{y_n\}$, and $\{\omega_n\}$ weakly converge to the same point in $zer(A+ B)$.

Note that the condition on the operator B in the multi-step inertial proximal contraction algorithm is weaker than that in the multi-step inertial forward–backward splitting method (6.2).

6.1 The Multi-step Inertial Krasnosel'skiĭ–Mann Iteration

Inspired by Liang's work, Dong et al. [65] introduced a multi-step inertial Krasnosel'skiĭ–Mann iteration.

Let $x_0, x_{-1} \in \mathscr{H}$ and, for each $n \geq 0$, set

$$
\begin{cases}
y_n = x_n + \displaystyle\sum_{k \in S_n} a_{n,k}(x_{n-k} - x_{n-k-1}), \\
z_n = x_n + \displaystyle\sum_{k \in S_n} b_{n,k}(x_{n-k} - x_{n-k-1}), \\
x_{n+1} = (1 - \lambda_n)y_n + \lambda_n T(z_n),
\end{cases}
\tag{6.5}
$$

where $S_n \subseteq \{0, 1, 2, \cdots, n-1\}$, $a_{n,k}, b_{n,k} \in (-1, 2]^{|S_n|}$ for each $n \geq 0$, and $|S_n|$ denotes the number of elements of the set S_n.

Remark 6.1 We review the relation of the multi-step inertial Krasnosel'skiĭ–Mann iteration to previous work. The multi-step inertial Krasnosel'skiĭ–Mann iteration is the most general inertial Krasnosel'skiĭ–Mann iteration and it is brand new to the literature while $\{0\} \subset S_n$.

For the case $S_n \equiv \{0\}$, it becomes the general inertial Krasnosel'skiĭ–Mann iteration (5.6) and recovers Krasnosel'skiĭ–Mann algorithm (3.1) if $S_n \equiv \emptyset$. For simplicity, set $a_n = a_{n,k}$ and $b_n = b_{n,k}$ when $S_n \equiv \{0\}$.

If we set $S_n \equiv \{0\}$, then, based on the choice of the inertial parameters a_n and b_n in the general inertial Krasnosel'skiĭ–Mann iteration, the relations between the multi-step inertial Krasnosel'skiĭ–Mann iteration with the aforementioned work are as follows:

(1) If $a_n = 0$ and $b_n = 0$ for each $n \geq 1$, then is the *original Krasnosel'skiĭ–Mann iteration* (3.1);

(2) If $a_n \in [0, a]$ and $b_n = 0$ for each $n \geq 1$, then this is the *reflected Krasnosel'skiĭ–Mann iteration* (5.9);

(3) If $a_n = 0$ and $b_n \in [0, b]$ for each $n \geq 1$, then this is the *accelerated Krasnosel'skiĭ–Mann iteration* (5.7);

(4) If $a_n \in [0, a]$ and $b_n = a_n$ for each $n \geq 1$, then this corresponds to the *classical inertial Krasnosel'skiĭ–Mann iteration* (5.5);

(5) If $a_n \in [0, a]$ and $b_n \in [0, b]$ with $a_n \neq b_n$ for each $n \geq 1$, then this is the *general inertial Krasnosel'skiĭ–Mann iteration* (5.6).

To show the convergence of the multi-step inertial Krasnosel'skiĭ–Mann iteration (6.5), we consider its relation with Krasnosel'skiĭ–Mann iteration with perturbations:

$$x_{n+1} = (1 - \lambda_n)(x_n + e_n^1) + \lambda_n T(x_n + e_n^2) \quad \text{for each } n \geq 0. \tag{6.6}$$

For each $n \geq 0$, set

$$e_n^1 = \sum_{k \in S_n} a_{n,k}(x_{n-k} - x_{n-k-1}) \tag{6.7}$$

and

$$e_n^2 = \sum_{k \in S_n} b_{n,k}(x_{n-k} - x_{n-k-1}), \tag{6.8}$$

then the algorithm (6.6) becomes the multi-step inertial Krasnosel'skiĭ–Mann iteration (6.5). So, we can get the convergence of the multi-step inertial Krasnosel'skiĭ–Mann iteration by using Theorem 3.9.

Theorem 6.2 *Let* $T : \mathscr{H} \to \mathscr{H}$ *be a nonexpansive operator with* $Fix(T) \neq \emptyset$. *Assume that* $\sum_{n=0}^{\infty} \lambda_n(1 - \lambda_n) = \infty$ *and*

$$\sum_{n=0}^{\infty} \max\{\max_{k \in S_n} |a_{n,k}|, \max_{k \in S_n} |b_{n,k}|\} \sum_{k \in S_n} \|x_{n-k} - x_{n-k-1}\| < \infty. \tag{6.9}$$

Then the sequence $\{x_n\}$ *generated by the multi-step inertial Krasnosel'skiĭ–Mann iteration* (6.5) *weakly converges to a point in* $Fix(T)$.

Remark 6.2 If the inertial parameters $a_{n,k}$ and $b_{n,k}$ are chosen in $[0, 1]$, then the condition (6.9) is simplified to

$$\sum_{n=0.}^{+\infty} \max\{\max_{k \in S_n} a_{n,k}, \max_{k \in S_n} b_{n,k}\} \sum_{k \in S_n} \|x_{n-k} - x_{n-k-1}\| < +\infty. \tag{6.10}$$

The condition (6.10) can be enforced by a simple online updating rule such as, for each $k \in S_n$ and $a_k, b_k \in [0, 1]$,

$$a_{n,k} = \min\{a_k, c_{a,n,k}\}, \quad b_{n,k} = \min\{b_k, c_{b,n,k}\}, \tag{6.11}$$

where $c_{a,n,k}, c_{b,n,k} > 0$ and $\max\{c_{a,n,k}, c_{b,n,k}\} \sum_{k \in S_n} \|x_{n-k} - x_{n-k-1}\|$ is summable. For instance, one can choose

$$c_{a,n,k} = \frac{c_{a,k}}{n^{1+\delta} \sum_{k \in S_n} \|x_{n-k} - x_{n-k-1}\|}, \quad c_{a,k} > 0, \quad \delta > 0$$

and, similarly, for $c_{b,k}$. In practice, most of the time, with proper choice of each a_n and b_n, (6.11) may never be triggered.

It is obvious that inertial parameter sequences $\{a_{n,k}\}_{n\in\mathbb{N},k\in S_n}$ and $\{b_{n,k}\}_{n\in\mathbb{N},k\in S_n}$ given by Remark 6.2 involve in the sequence $\{x_n\}$.

Remark 6.3

(1) According to Theorem 3.14, to get the pointwise convergence rate of $O(1/\sqrt{n})$ for the multi-step inertial Krasnosel'skiĭ–Mann iteration, we need $0 < \inf_{n\in\mathbb{N}} \lambda_n \le \sup_{n\in\mathbb{N}} \lambda_n < 1$ and

$$\sum_{n=0}^{+\infty}(n+1)\max\{\max_{k\in S_n}|a_{n,k}|, \max_{k\in S_n}|b_{n,k}|\}\sum_{k\in S_n}\|x_{n-k}-x_{n-k-1}\| < +\infty.$$

One can choose inertial parameters $\{a_{n,k}\}_{n\in\mathbb{N},k\in S_n}$ and $\{b_{n,k}\}_{n\in\mathbb{N},k\in S_n}$ by the same online updating rule as Remark 6.2.

(2) According to Theorem 3.15, to obtain the ergodic convergence rate of the multi-step inertial Krasnosel'skiĭ–Mann iteration, one can similarly choose inertial parameters $\{a_{n,k}\}_{n\in\mathbb{N},k\in S_n}$ and $\{b_{n,k}\}_{n\in\mathbb{N},k\in S_n}$.

When $S_n \equiv \{0\}$, the multi-step inertial Krasnosel'skiĭ–Mann iteration reduces to the general inertial Krasnosel'skiĭ–Mann algorithm (5.6) and then we have the following theorem from Theorem 6.2. To lighten the notations, for $S_n \equiv \{0\}$, we denote $a_n = a_{n,0}$ and $b_n = b_{n,0}$ for each $n \ge 1$.

Theorem 6.3 (The Conditional Convergence for $S_n \equiv \{0\}$) *Let $T : \mathcal{H} \to \mathcal{H}$ be a nonexpansive operator with $Fix(T) \ne \emptyset$. Assume that $\sum_{n=0}^{\infty} \lambda_n(1-\lambda_n) = \infty$ and*

$$\sum_{n=0}^{+\infty}\max\{a_n, b_n\}\|x_n - x_{n-1}\| < \infty. \tag{6.12}$$

Then the sequence $\{x_n\}$ generated by the general inertial Krasnosel'skiĭ–Mann algorithm (5.6) weakly converges to a point in $Fix(T)$.

The inertial parameters in Theorem 6.3 depend on the sequence $\{x_n\}_{n\in\mathbb{N}}$ and therefore Theorem 6.3 can be seen as the conditional convergence. In contrast to this, the inertial parameters in Theorem 5.2 do not involve the sequence $\{x_n\}_{n\in\mathbb{N}}$ and therefore Theorem 5.2 can be seen as the unconditional convergence.

Remark 6.4 When $b_n = a_n$ for each $n \geq 1$, then the condition (6.12) becomes

$$\sum_{n=0}^{+\infty} a_n \|x_n - x_{n-1}\| < +\infty,$$

which is different from the condition (C2).

It is easy to see that if $|S_n|$ is large, for example, $S_n = \{0, 1, \cdots, n-1\}$ and $|S_n| = n$, then the computational cost of $a_{n,k}$ and $b_{n,k}$ by using (6.11) may be very expensive since one needs to compute the inertial parameters per iteration. This observation motivates us to seek inertial parameters that do not depend on the iterative sequence $\{x_n\}$.

6.2 Two Inertial Parameter Sequences That Do Not Depend on the Iterative Sequence

In this section, we present two inertial parameter sequences of the multi-step inertial Krasnosel'skiĭ–Mann iteration (6.5) that do not involve the iterative sequence $\{x_n\}$.

Very recently, Dong et al. [67] extended the algorithm (3.5) and considered a general case on the affine hull of orbits.

(A) The general Krasnosel'skiĭ–Mann iteration on the affine hull of orbits

Let $\lambda \in (0, 1)$, and let $\{\mu_{n,k}\}_{n\in\mathbb{N},0\leq k\leq n}$ and $\{\nu_{n,k}\}_{n\in\mathbb{N},0\leq k\leq n}$ be two real arrays satisfying the assumptions (B1)–(B4). Take $x_0 \in \mathscr{H}$ and, for each $n \geq 0$, define

$$\begin{cases} y_n = \sum_{k=0}^{n} \mu_{n,k} x_k, \\ z_n = \sum_{k=0}^{n} \nu_{n,k} x_k, \\ x_{n+1} = (1-\lambda) y_n + \lambda T(z_n). \end{cases} \tag{6.13}$$

Remark 6.5 The relations between the algorithm (6.13) and other related work are as follows:

(1) The scheme (6.13) can be seen as a special case of the multi-step inertial Krasnosel'skiĭ–Mann iteration (6.5). Indeed, by the first formula of (6.13), we have

$$y_n = \sum_{k=0}^{n} \mu_{n,k} x_k$$

$$= x_n + (\mu_{n,n} - 1)(x_n - x_{n-1}) + (\mu_{n,n} + \mu_{n,n-1} - 1)(x_{n-1} - x_{n-2})$$

$$+ \cdots + \left(\sum_{k=1}^{n} \mu_{n,k} - 1\right)(x_1 - x_0) + \left(\sum_{k=0}^{n} \mu_{n,k} - 1\right) x_0$$

$$= x_n + (\mu_{n,n} - 1)(x_n - x_{n-1}) + (\mu_{n,n} + \mu_{n,n-1} - 1)(x_{n-1} - x_{n-2})$$

$$+ \cdots + \left(\sum_{k=1}^{n} \mu_{n,k} - 1\right)(x_1 - x_0),$$

$$(6.14)$$

where the last equality comes from (B2). Similarly, we get

$$z_n = x_n + (\nu_{n,n} - 1)(x_n - x_{n-1}) + (\nu_{n,n} + \nu_{n,n-1} - 1)(x_{n-1} - x_{n-2})$$

$$+ \cdots + \left(\sum_{k=1}^{n} \nu_{n,k} - 1\right)(x_1 - x_0).$$

$$(6.15)$$

Let $S_n = \{0, 1, \cdots, n-1\}$, $a_{n,k} = \sum_{j=n-k}^{n} \mu_{n,j} - 1$, $b_{n,k} = \sum_{j=n-k}^{n} \nu_{n,j} - 1$, and $\lambda_n \equiv \lambda$ for each $n \geq 0$ and $0 \leq k \leq n$, then the multi-step inertial Krasnosel'skiĭ–Mann iteration (6.5) becomes the scheme (6.13).

(2) We list three special cases of the scheme (6.13):

(a) Let $\nu_{n,n} = 1$ and $\nu_{n,k} = 0$ for each $n \geq 1$ and $0 \leq k \leq n - 1$, then the scheme (6.13) reduces to the following:

$$\begin{cases} y_n = \displaystyle\sum_{k=0}^{n} \mu_{n,k} x_k, \\ \\ x_{n+1} = (1 - \lambda) y_n + \lambda T(x_n). \end{cases}$$

$$(6.16)$$

(b) Let $\mu_{n,n} = 1$ and $\mu_{n,k} = 0$ for each $n \geq 1$ and $0 \leq k \leq n - 1$, then the scheme (6.13) reduces to the following:

$$\begin{cases} z_n = \displaystyle\sum_{k=0}^{n} \nu_{n,k} x_k, \\ \\ x_{n+1} = (1 - \lambda) x_n + \lambda T(z_n). \end{cases}$$

$$(6.17)$$

(c) Let $\nu_{n,k} = \mu_{n,k}$ for each $n \geq 1$ and $0 \leq k \leq n$, then the scheme (6.13) reduces to (3.5) with $\lambda_n \equiv \lambda$ and $a_{nk} = \mu_{n,k}$ for each $n \geq 1$ and $0 \leq k \leq n$.

The convergence result of the scheme (6.13) is given as follows:

Theorem 6.4 ([67, Theorem 3.1]) *Let $T : \mathcal{H} \to \mathcal{H}$ be a nonexpansive mapping with FixT $\neq \emptyset$. Assume that the sequences $\{\mu_{n,k}\}_{n\in\mathbb{N},0\leq k\leq n}$, $\{\nu_{n,k}\}_{n\in\mathbb{N},0\leq k\leq n}$, and $\{(1-\lambda)\mu_{n,k}+\lambda\nu_{n,k}\}_{n\in\mathbb{N},0\leq k\leq n}$ for all $\lambda \in (0, 1)$ satisfy (B1)–(B4) with the sequence $\{\chi_n\}$. Let $\{x_n\}$ generated by the algorithm (6.13) satisfy*

$$\sum_{n=0}^{+\infty} \chi_n \sum_{k=0}^{n} \sum_{j=0}^{n} \left((1 - \lambda)[\mu_{n,k}\mu_{n,j}]^- + \lambda[\nu_{n,k}\nu_{n,j}]^-\right) \|x_k-x_j\|^2 < +\infty. \quad (6.18)$$

Then the sequence $\{x_n\}$ converges weakly to a point in Fix(T).

The condition (6.18) seems difficult to verify since it involves the parameters $\{\mu_{n,k}\}_{n\in\mathbb{N},0\leq k\leq n}$ and $\{\nu_{n,k}\}_{n\in\mathbb{N},0\leq k\leq n}$, the sequence $\{\chi_n\}$, and the iterative sequence $\{x_n\}$. However, the condition (6.18) can be got rid of when $\{\mu_{n,k}\}_{n\in\mathbb{N},0\leq k\leq n}$ and $\{\nu_{n,k}\}_{n\in\mathbb{N},0\leq k\leq n}$ are nonnegative.

Corollary 6.1 *Let $T : \mathcal{H} \to \mathcal{H}$ be a nonexpansive mapping with FixT $\neq \emptyset$. Assume that $\{\mu_{n,k}\}_{n\in\mathbb{N},0\leq k\leq n}$ and $\{\nu_{n,k}\}_{n\in\mathbb{N},0\leq k\leq n}$ are two nonnegative sequences such that the sequences $\{\mu_{n,k}\}_{n\in\mathbb{N},0\leq k\leq n}$, $\{\nu_{n,k}\}_{n\in\mathbb{N},0\leq k\leq n}$, and $\{(1 - \lambda)\mu_{n,k} + \lambda\nu_{n,k}\}_{n\in\mathbb{N},0\leq k\leq n}$ for all $\lambda \in (0, 1)$ satisfy (A1)–(A3) and (A5). Then the sequence $\{x_n\}$ generated by the algorithm (6.13) converges weakly to a point in Fix(T).*

By using Example 3.2, we present two nonnegative arrays $\{\mu_{n,k}\}_{n\in\mathbb{N},0\leq k\leq n}$ and $\{\nu_{n,k}\}_{n\in\mathbb{N},0\leq k\leq n}$ satisfying (A1)–(A3) and (A5).

Example 6.1 Let $p, q \geq 1$. Take the sequences $\{\mu_i\}_{0\leq i\leq p}$ and $\{\nu_j\}_{0\leq j\leq q}$ of strictly positive numbers such that $\sum_{i=0}^{p} \mu_i = 1$ and $\sum_{j=0}^{q} \nu_j = 1$, respectively. Define the sequence $\{\mu_{n,k}\}_{n\in\mathbb{N},0\leq k\leq n}$ as follows:

for $p = 0$, $\mu_{n,n} = \mu_0 = 1$ and $\mu_{n,k} \equiv 0$ for each $n \geq 1$ and $0 \leq k < n$;

for each $p \geq 1$, for each $n \in \{0, \cdots, p - 1\}$,

$$\mu_{n,k} = \begin{cases} 0, & \text{if } 0 \leq k < n, \\ 1, & \text{if } k = n; \end{cases}$$

for each $n \geq p$,

$$\mu_{n,k} = \begin{cases} 0, & \text{if } 0 \leq k < n - p, \\ \mu_{n-k}, & \text{if } n - p \leq k \leq n. \end{cases}$$

Define the sequence $\{\nu_{n,k}\}_{n\in\mathbb{N},0\leq k\leq n}$ as follows:

for $q = 0$, $\nu_{n,n} = \nu_0 = 1$ and $\nu_{n,k} \equiv 0$ for each $n \geq 1$ and $0 \leq k < n$;

for each $q \geq 1$, for each $n \in \{0, \cdots, q - 1\}$,

$$\nu_{n,k} = \begin{cases} 0, & \text{if } 0 \leq k < n, \\ 1, & \text{if } k = n; \end{cases}$$

for each $n \geq p$,

$$v_{n,k} = \begin{cases} 0, & \text{if } 0 \leq k < n - p, \\ v_{n-k}, & \text{if } n - p \leq k \leq n. \end{cases}$$

Remark 6.6 We give some illustrations for Example 6.1 as follows:

(1) The arrays $\{\mu_{n,k}\}_{n \in \mathbb{N}, 0 \leq k \leq n}$ and $\{v_{n,k}\}_{n \in \mathbb{N}, 0 \leq k \leq n}$ are not exactly the same as Example 3.2 since they include the special cases that $\mu_{n,n} \equiv 1$ and $\mu_{n,k} \equiv 0$ for $p = 0$ and $v_{n,n} \equiv 1$ and $v_{n,k} \equiv 0$ for $q = 0$. These two special cases correspond to algorithms (6.17) and (6.16), respectively.

(2) By using [46, Example 2.6], we know that the sequences $\{\mu_{n,k}\}_{n \in \mathbb{N}, 0 \leq k \leq n}$, $\{v_{n,k}\}_{n \in \mathbb{N}, 0 \leq k \leq n}$, and $\{(1 - \lambda)\mu_{n,k} + \lambda v_{n,k}\}_{n \in \mathbb{N}, 0 \leq k \leq n}$ for all $\lambda \in (0, 1)$ satisfy (A1)–(A3) and (A5).

(3) Using Example 6.1, one can construct the (p, q)-step inertial Krasnosel'skiĭ–Mann iteration (denoted by (p, q)-MiKM), in which the computations of y_n and z_n involve $\{x_n, \cdots, x_{n-p}\}$ and $\{x_n, \cdots, x_{n-q}\}$, respectively. By (2), the (p, q)-MiKM weakly converges to a point in Fix$T$, provided that $T : \mathcal{H} \to \mathcal{H}$ is a nonexpansive mapping with Fix$T \neq \emptyset$.

By using Example 3.1, we give another example, where $\{\mu_{n,k}\}_{n \in \mathbb{N}, 0 \leq k \leq n}$ and $\{v_{n,k}\}_{n \in \mathbb{N}, 0 \leq k \leq n}$ were shown to satisfy the conditions (A1)–(A3) and (A5) in (see [46, Example 2.5]).

Example 6.2 Let $\mu_{n,n} = 1 - \mu$ and $v_{n,n} = 1 - v$ for each $n \geq 0$, where $\mu, v \in (0, 1)$. Take

$$\mu_{n,k} = \begin{cases} \mu^n, & \text{if } k = 0, \\ (1 - \mu)\mu^{n-k}, & \text{if } 1 \leq k \leq n, \end{cases}$$

and

$$v_{n,k} = \begin{cases} v^n, & \text{if } k = 0, \\ (1 - v)v^{n-k}, & \text{if } 1 \leq k \leq n. \end{cases}$$

Remark 6.7 We give two illustrations for Example 6.2 as follows:

(1) Although the sequences $\{\mu_{n,k}\}_{n \in \mathbb{N}, 0 \leq k \leq n}$ and $\{v_{n,k}\}_{n \in \mathbb{N}, 0 \leq k \leq n}$ satisfy (A1)–(A3) and (A5), whether the sequence $\{(1 - \lambda)\mu_{n,k} + \lambda v_{n,k}\}_{n \in \mathbb{N}, 0 \leq k \leq n}$ for all $\lambda \in (0, 1)$ satisfies (A1)–(A3) and (A5) is unknown.

(2) For the algorithm (6.13) with arrays $\{\mu_{n,k}\}_{n \in \mathbb{N}, 0 \leq k \leq n}$ and $\{v_{n,k}\}_{n \in \mathbb{N}, 0 \leq k \leq n}$ given in Example 6.2, the computation of x_{n+1} involves $\{x_n, x_{n-1}, \cdots, x_0\}$ since the sequences $\{\mu_{n,k}\}_{n \in \mathbb{N}, 0 \leq k \leq n}$ and $\{v_{n,k}\}_{n \in \mathbb{N}, 0 \leq k \leq n}$ are strictly positive arrays.

By Remark 6.7 (1), the convergence of the algorithm (6.13) with the arrays $\{\mu_{n,k}\}_{n \in \mathbb{N}, 0 \leq k \leq n}$ and $\{v_{n,k}\}_{n \in \mathbb{N}, 0 \leq k \leq n}$ given in Example 6.2 is unknown.

To verify it, we introduce the following modified Krasnosel'skiĭ–Mann iteration.

(B) The modified Krasnosel'skiĭ–Mann iteration

For any $x_0, y_0, z_0 \in \mathscr{H}$ and $n \geq 1$, calculate

$$\begin{cases} y_{n+1} = (1 - \alpha)x_n + \alpha y_n, \\ z_{n+1} = (1 - \beta)x_n + \beta z_n, \\ x_{n+1} = (1 - \lambda)y_{n+1} + \lambda T(z_{n+1}), \end{cases} \tag{6.19}$$

where $\lambda \in (0, 1)$ and $\alpha, \beta \in [0, 1)$.

Remark 6.8 Below we highlight the relation of the algorithm (6.19) and the algorithm (6.13) and give some special cases of the algorithm (6.19).

(1) If $\alpha, \beta \in (0, 1)$, $y_0 = x_0$, and $z_0 = x_0$, then the algorithm (6.19) becomes a special case of the algorithm (6.13). Indeed, set $y_0 = x_0$ in Algorithm (6.19). Repeating $y_{n+1} = (1 - \alpha)x_n + \alpha y_n$, we have

$$y_{n+1} = (1 - \alpha)x_n + \alpha y_n$$

$$= (1 - \alpha)x_n + \alpha \left[(1 - \alpha)x_{n-1} + \alpha y_{n-1}\right]$$

$$= \cdots$$

$$= \sum_{k=1}^{n}(1 - \alpha)\alpha^{n-k}x_k + \alpha^n x_0.$$

Similarly, set $z_0 = x_0$ and then we obtain

$$z_{n+1} = \sum_{k=1}^{n}(1 - \beta)\beta^{n-k}x_k + \beta^n x_0.$$

Let

$$\mu_{n,k} = \begin{cases} \alpha^n, & \text{if } k = 0, \\ (1 - \alpha)\alpha^{n-k}, & \text{if } 1 \leq k \leq n, \end{cases} \tag{6.20}$$

and

$$\nu_{n,k} = \begin{cases} \beta^n, & \text{if } k = 0, \\ (1 - \beta)\beta^{n-k}, & \text{if } 1 \leq k \leq n. \end{cases} \tag{6.21}$$

It is obvious that the arrays $\{\mu_{n,k}\}_{n \in \mathbb{N}, 0 \leq k \leq n}$ and $\{\nu_{n,k}\}_{n \in \mathbb{N}, 0 \leq k \leq n}$ in (6.20) and (6.21) are given as in Example 6.2.

(2) Based on the choice of the parameters α and β, we discuss some special cases of the algorithm (6.19).

(a) Letting $\beta = 0$ and $\lambda, \alpha \in (0, 1)$ in the algorithm (6.19), we get

$$\begin{cases} y_{n+1} = (1 - \alpha)x_n + \alpha y_n, \\ x_{n+1} = (1 - \lambda)y_{n+1} + \lambda T(x_n). \end{cases} \tag{6.22}$$

(b) Letting $\alpha = 0$ and $\lambda, \beta \in (0, 1)$ in the algorithm (6.19), we obtain

$$\begin{cases} z_{n+1} = (1 - \beta)x_n + \beta z_n, \\ x_{n+1} = (1 - \lambda)x_n + \lambda T(z_{n+1}). \end{cases}$$

(c) Letting $\lambda \in (0, 1)$ and setting $\beta = \alpha$ in the algorithm (6.19), we have

$$\begin{cases} y_{n+1} = (1 - \alpha)x_n + \alpha y_n, \\ x_{n+1} = (1 - \lambda)y_{n+1} + \lambda T(y_{n+1}). \end{cases}$$

(d) Letting $\alpha = \beta = 0$, the algorithm (6.19) reduces to the Krasnosel'skiĭ–Mann iteration (3.1) with $\lambda_n \equiv \lambda$.

Theorem 6.5 ([67, Theorem 4.1]) *Suppose that $T : \mathscr{H} \to \mathscr{H}$ is a nonexpansive mapping with $\mathrm{Fix}(T) \neq \emptyset$. Then the iterative sequence $\{x_n\}$ generated by the algorithm (6.19) weakly converges to a point in $\mathrm{Fix}(T)$.*

Next, we give a running-average iteration-complexity bound of the algorithm (6.19).

Theorem 6.6 ([67, Theorem 4.2]) *Suppose that $T : \mathscr{H} \to \mathscr{H}$ is a nonexpansive mapping with $\mathrm{Fix}(T) \neq \emptyset$. Let $\{x_n\}$ be the iterative sequence generated by the algorithm (6.19). Then we have*

$$\frac{1}{n} \sum_{k=0}^{n} \|x_k - T(x_k)\|^2 \le \frac{3}{n} \left(\frac{C}{\tau} + \|y_0 - T(z_0)\|^2 \right), \tag{6.23}$$

where

$$C = \inf_{x \in Fix(T)} \left\{ \|x_0 - x\|^2 + \frac{(1 - \lambda)\alpha}{1 - \alpha} \|y_0 - x\|^2 + \frac{\lambda \beta}{1 - \beta} \|z_0 - x\|^2 \right\}$$

and

$$\tau = \min \{\lambda(1 - \lambda), (1 - \lambda)\alpha, \lambda\beta\}.$$

Remark 6.9 The computation of x_{n+1} in the algorithm (6.19) just involves x_n, y_n, and z_n and does not need the information of $\{x_0, x_1, \cdots, x_{n-1}\}$. A similar conclusion is reached for the algorithm (6.13) with the arrays $\{\mu_{n,k}\}_{n\in\mathbb{N},0\le k\le n}$ and $\{\nu_{n,k}\}_{n\in\mathbb{N},0\le k\le n}$ given in Example 6.2, which is different from Remark 6.7 (2).

6.3   Some Applications

In this section, we introduce some multi-step inertial operator splitting methods by using the relations of the Krasnosel'skiĭ–Mann iteration and the operator splitting methods.

By combining the multi-step inertial Krasnosel'skiĭ–Mann iteration and the forward–backward splitting method, we propose the following algorithm.

(A) The multi-step inertial forward–backward splitting method

Let $A : \mathscr{H} \rightrightarrows \mathscr{H}$ be a maximally monotone operator and $B : \mathscr{H} \to \mathscr{H}$ be a β-cocoercive mapping for some $\beta > 0$. Let $\gamma \in (0, 2\beta)$ and $\{\lambda_n\}$ in $\left[0, \frac{4\beta-\gamma}{2\beta}\right]$ such that $\sum_{n=0}^{+\infty} \lambda_n \left(\frac{4\beta-\gamma}{2\beta} - \lambda_n\right) = +\infty$. Choose $x_0, x_{-1} \in \mathscr{H}$ arbitrarily and, for each $n \ge 0$, apply the following iteration:

$$\begin{cases} y_n = x_n + \displaystyle\sum_{k\in S_n} a_{n,k}(x_{n-k} - x_{n-k-1}), \\[2mm] z_n = x_n + \displaystyle\sum_{k\in S_n} b_{n,k}(x_{n-k} - x_{n-k-1}), \\[2mm] x_{n+1} = (1-\lambda_n)y_n + \lambda_n J_{\gamma A}(\mathrm{Id} - \gamma B)z_n. \end{cases} \qquad (6.24)$$

Combining Theorem 6.2 and [15, Theorem 26.14], we present the convergence of the multi-step inertial forward–backward splitting method (6.24).

Theorem 6.7 *Let $A : \mathscr{H} \rightrightarrows \mathscr{H}$ be maximally monotone and $B : \mathscr{H} \to \mathscr{H}$ be β-cocoercive for some $\beta \in (0, +\infty)$. Let $\gamma \in (0, 2\beta)$ and set $\delta = 2 - \gamma/(2\beta)$. Furthermore, let $\{\lambda_n\}$ be the sequence in $[0, \delta]$ such that*

$$\sum_{n=0}^{+\infty} \lambda_n(\delta - \lambda_n) = +\infty.$$

Assume that

$$\sum_{n=0}^{+\infty} \max\{\max_{k\in S_n}|a_{n,k}|, \max_{k\in S_n}|b_{n,k}|\} \sum_{k\in S_n} \|x_{n-k} - x_{n-k-1}\| < +\infty. \qquad (6.25)$$

Let $x_0, x_{-1} \in \mathcal{H}$ and suppose that $zer(A + B) \neq \emptyset$. Then the sequence $\{x_n\}$ generated by the multi-step inertial forward–backward splitting method (6.24) weakly converges to a point in $zer(A + B)$.

It is easy to verify that the algorithm (6.24) is different from the algorithm (6.2) presented by Liang [125]. Their convergence conditions are also different.

Remark 6.10 We illustrate two differences of the sets S and S_k in the algorithm (6.24) and the algorithm (6.2):

(1) Note that the number of the elements of set $S = \{0, 1, \cdots, s - 1\}$ in the algorithm (6.2) is s. So, to generate the sequence $\{x_n\}$, $\{x_{-1}, x_{-2}, \cdots, x_{-s}\}$ must be determined. Ortega and Rheinboldt [164] gave the specification of certain sets of the points $\{x_{-1}, x_{-2}, \cdots, x_{-s}\}$. Liang [125] assumed that $x_{-k} = x_0$ for each $k \in \{1, 2, \cdots, s\}$. While in the algorithm (6.24), only two initial points x_0, x_{-1} are needed.
(2) For each $n \geq s$, the set $S = \{0, 1, \cdots, s - 1\}$ is a fixed subset of $\{0, 1, \cdots, n - 1\}$, while S_n can be any subset of $\{0, 1, \cdots, n - 1\}$. It is obvious that S_n are more flexible than S.

In fact, the choice of the set S_n is more reasonable since the computation of x_{n+1} involves $x_0, x_1, \cdots, x_n$ for the algorithm (6.13) with the arrays $\{\mu_{n,k}\}_{n \in \mathbb{N}, 0 \leq k \leq n}$ and $\{\nu_{n,k}\}_{n \in \mathbb{N}, 0 \leq k \leq n}$ given in Example 6.1.

Next, we present the multi-step inertial backward–forward splitting method as follows.

(B) The multi-step inertial backward–forward splitting method

Let $A : \mathcal{H} \rightrightarrows \mathcal{H}$ be maximally monotone and $B : \mathcal{H} \to \mathcal{H}$ be β-cocoercive for some $\beta > 0$. Let $\gamma \in (0, 2\beta)$ and $\{\lambda_n\}$ in $\left[0, \frac{4\beta - \gamma}{2\beta}\right]$ such that

$$\sum_{n=0}^{+\infty} \lambda_n \left(\frac{4\beta - \gamma}{2\beta} - \lambda_n\right) = +\infty.$$

Choose $x_0, x_{-1} \in \mathcal{H}$ arbitrarily and, for each $n \geq 0$, apply the following iteration:

$$\begin{cases} u_n = x_n + \sum_{k \in S_n} a_{n,k}(x_{n-k} - x_{n-k-1}), \\[2mm] v_n = x_n + \sum_{k \in S_n} b_{n,k}(x_{n-k} - x_{n-k-1}), \\[2mm] y_n = J_{\gamma A}(v_n), \\[2mm] z_n = y_n - \gamma B y_n, \\[2mm] x_{n+1} = u_n + \lambda_n(z_n - u_n). \end{cases} \qquad (6.26)$$

From Theorem 6.7 and [5, Theorem 3.5], we present the convergence of the multi-step inertial backward–forward splitting method (6.26).

Theorem 6.8 *Let $A : \mathcal{H} \rightrightarrows \mathcal{H}$ be maximally monotone and $B : \mathcal{H} \to \mathcal{H}$ be β-cocoercive for some $\beta \in (0, +\infty)$. Let $\gamma \in (0, 2\beta)$ and set $\delta = \min\{1, \beta/\gamma\} + 1/2$. Furthermore, let $\{\lambda_n\}$ be the sequence in $[0, \delta]$ such that*

$$\sum_{n=0}^{+\infty} \lambda_n(\delta - \lambda_n) = +\infty.$$

Assume that

$$\sum_{n=0}^{+\infty} \max\{\max_{k \in S_n} |a_{n,k}|, \max_{k \in S_n} |b_{n,k}|\} \sum_{k \in S_n} \|x_{n-k} - x_{n-k-1}\| < +\infty.$$

Let $x_0, x_{-1} \in \mathcal{H}$ and suppose that $zer(A + B) \neq \emptyset$. Let $\{y_n\}$ be the sequence generated by the multi-step inertial backward–forward splitting method (6.26). Then the sequence $\{y_n\}$ weakly converges to a point in $zer(A + B)$.

By combining the multi-step inertial Krasnosel'skiĭ–Mann iteration and the primal–dual splitting method, we propose the following algorithm.

(C) The multi-step inertial primal–dual splitting method

Let $A : \mathcal{H} \rightrightarrows \mathcal{H}$ be maximally monotone, $B : \mathcal{H} \to \mathcal{H}$ be β_B-cocoercive for some $\beta_B > 0$, $C : \mathcal{H} \rightrightarrows \mathcal{H}$ be maximally monotone, and, moreover, D be β_D-strongly monotone for some $\beta_D > 0$, $L : \mathcal{H} \to \mathcal{G}$ be a bounded linear operator. Choose $\gamma_A, \gamma_C > 0$ such that

$$2 \min\{\beta_B, \beta_D\} \min\left\{\frac{1}{\gamma_A}, \frac{1}{\gamma_C}\right\}\left(1 - \sqrt{\gamma_A \gamma_C \|L\|^2}\right) > 1. \tag{6.27}$$

Choose $x_0, x_{-1} \in \mathcal{H}$ and $v_0, v_{-1} \in \mathcal{G}$ arbitrarily and, for each $n \geq 0$, apply the following iteration:

$$\begin{cases} y_n = x_n + \sum_{k \in S_n} a_{n,k}(x_{n-k} - x_{n-k-1}), \\[2mm] u_n = x_n + \sum_{k \in S_n} b_{n,k}(v_{n-k} - v_{n-k-1}), \\[2mm] x_{n+1} = J_{\gamma_A A}(y_n - \alpha B(y_n) - \gamma_A L^* u_n), \\[2mm] \bar{x}_{n+1} = 2x_{n+1} - y_n, \\[2mm] v_{n+1} = J_{\gamma_C C^{-1}}\left(u_n - \gamma_C D^{-1}(u_n) + \gamma_C L\bar{x}_{n+1}\right). \end{cases} \tag{6.28}$$

Following Theorem 6.2, we get the convergence of the multi-step inertial primal–dual splitting method (6.28).

Theorem 6.9 *Suppose that the condition (4.18) holds. Assume that*

$$\sum_{n=0}^{+\infty} \max_{k \in S_n} |a_{n,k}| \sum_{k \in S_n} \|x_{n-k} - x_{n-k-1}\| < +\infty,$$

and

$$\sum_{n=0}^{+\infty} \max_{k \in S_n} |b_{n,k}| \sum_{k \in S_n} \|v_{n-k} - v_{n-k-1}\| < +\infty.$$

Let x_0, $x_{-1} \in \mathcal{H}$ and v_0, $v_{-1} \in \mathcal{G}$. Then the sequences $\{x_n\}$ and $\{v_n\}$ generated by the multi-step inertial primal–dual splitting method (6.28) weakly converges to a point in $\mathcal{P}$ and $\mathcal{D}$, respectively.

Remark 6.11 Liang [125] presented a more general multi-step inertial primal–dual splitting method as follows: for each $n \geq 0$,

$$
\begin{cases}
y_{n,1} = x_n + \sum_{k \in S} a_{n,k}(x_{n-k} - x_{n-k-1}), \\[2mm]
y_{n,2} = x_n + \sum_{k \in S} b_{n,k}(x_{n-k} - x_{n-k-1}), \\[2mm]
u_{n,1} = x_n + \sum_{k \in S} a_{n,k}(v_{n-k} - v_{n-k-1}), \\[2mm]
u_{n,2} = x_n + \sum_{k \in S} b_{n,k}(v_{n-k} - v_{n-k-1}), \\[2mm]
x_{n+1} = J_{\alpha A}(y_{n,1} - \alpha B(y_{n,2}) - \alpha L^* u_{n,1}), \\[2mm]
\bar{x}_{n+1} = 2x_{n+1} - y_{n,1}, \\[2mm]
v_{n+1} = J_{\beta C^{-1}}\left(u_{n,1} - \beta D^{-1}(u_{n,2}) + \beta L \bar{x}_{n+1}\right).
\end{cases}
\tag{6.29}
$$

Note that the algorithm (6.28) is a special case of the algorithm (6.29) if setting $S_n = S$, but the convergence condition is different.

By combining the multi-step inertial Krasnosel'skiĭ–Mann iteration and the Douglas–Rachford splitting method, we propose the following method.

(D) The multi-step inertial Douglas–Rachford splitting method

Let A_1, $A_2 : \mathcal{H} \rightrightarrows \mathcal{H}$ be two maximally monotone operators. Let $\gamma > 0$ and $\{\lambda_n\}$ be the sequence in $[0, 2]$ such that

$$\sum_{n=0}^{+\infty} \lambda_n(2 - \lambda_n) = +\infty.$$

Choose $x_0, x_{-1} \in \mathcal{H}$ arbitrarily and, for each $n \geq 0$, apply the following iteration:

$$
\begin{cases}
u_n = x_n + \sum_{k \in S_n} a_{n,k}(x_{n-k} - x_{n-k-1}), \\
\\
v_n = x_n + \sum_{k \in S_n} b_{n,k}(x_{n-k} - x_{n-k-1}), \\
\\
x_{n+1} = \left(1 - \dfrac{\lambda_n}{2}\right) u_n + \dfrac{\lambda_n}{2} R_{\gamma A_1} R_{\gamma A_2}(v_n).
\end{cases}
\tag{6.30}
$$

From Theorem 6.2 and the nonexpansivity of $R_{\gamma A_1} R_{\gamma A_2}$, we get the convergence of the multi-step inertial Douglas–Rachford splitting method (6.30).

Theorem 6.10 *Assume that $\lambda_n \in [0, 2]$ satisfies*

$$\sum_{n=0}^{+\infty} \lambda_n(2 - \lambda_n) = +\infty$$

and

$$\sum_{n=0}^{+\infty} \max\{\max_{k \in S_n} |a_{n,k}|, \max_{k \in S_n} |b_{n,k}|\} \sum_{k \in S_n} \|x_{n-k} - x_{n-k-1}\| < +\infty.$$

Then the sequence $\{x_n\}$ generated by the multi-step inertial Douglas–Rachford splitting method (6.30) weakly converges to a point in $\mathrm{Fix}(R_{\gamma A_1} R_{\gamma A_2})$.

Letting $b_{n,k} = a_{n,k}$ for each $n \geq 1$ and $0 \leq k \leq n$ in (6.30), we get a special case of the multi-step inertial Douglas–Rachford splitting method (6.30) as follows: for each $n \geq 0$,

$$
\begin{cases}
u_n = x_n + \sum_{k \in S_n} a_{n,k}(x_{n-k} - x_{n-k-1}), \\
\\
y_n = J_{\gamma A_2} u_n, \\
\\
z_n = J_{\gamma A_1}(2y_n - u_n), \\
\\
x_{n+1} = \left(1 - \dfrac{\lambda_n}{2}\right) u_n + \lambda_n \left(z_n - y_n + \dfrac{1}{2} u_n\right).
\end{cases}
\tag{6.31}
$$

Note that when $\lambda_n \equiv 1$ and $S_n = S$, the algorithm (6.31) degenerates [125, Algorithm 8].

By combining the multi-step inertial Krasnosel'skiĭ–Mann iteration and Davis–Yin splitting method, we propose the following method.

(E) The multi-step inertial Davis–Yin splitting method

Let $A_1, A_2 : \mathscr{H} \rightrightarrows \mathscr{H}$ be maximally monotone and $B : \mathscr{H} \to \mathscr{H}$ be β-cocoercive for some $\beta > 0$. Let $\gamma \in (0, 2\beta)$ and $\{\lambda_n\}$ be the sequence in $\left[0, \frac{4\beta - \gamma}{2\beta}\right]$ such that

$$\sum_{n=0}^{+\infty} \lambda_n \left(\frac{4\beta - \gamma}{2\beta} - \lambda_n \right) = +\infty.$$

Choose $x_0, x_{-1} \in \mathscr{H}$ arbitrarily and, for each $n \geq 0$, apply the following iteration:

$$\begin{cases} u_n = x_n + \sum_{k \in S_n} a_{n,k}(x_{n-k} - x_{n-k-1}), \\[2mm] v_n = x_n + \sum_{k \in S_n} b_{n,k}(x_{n-k} - x_{n-k-1}), \\[2mm] y_n = J_{\gamma A_2} v_n, \\[2mm] z_n = J_{\gamma A_1}(2y_n - v_n - \gamma B y_n), \\[2mm] x_{n+1} = (1 - \lambda_n) u_n + \lambda_k (z_n - y_n + v_n). \end{cases} \qquad (6.32)$$

Combining Theorem 6.2 and [55, Theorem 2.1], we present the convergence of the multi-step inertial Davis–Yin splitting method (6.32).

Theorem 6.11 *Set* $\gamma \in (0, 2\beta\varepsilon)$, *where* $\varepsilon \in (0, 1)$. *Assume that* $\lambda_n \subseteq (0, 1/\alpha)$, *where* $\alpha = 1/(2 - \varepsilon) < 2\beta/(4\beta - \gamma)$ *satisfies*

$$\sum_{n=0}^{+\infty} \left(1 - \frac{\lambda_n}{\alpha}\right) \frac{\lambda_n}{\alpha} = +\infty,$$

and

$$\sum_{n=0}^{+\infty} \max_{k \in S_n} |a_{n,k}| \sum_{k \in S_n} \|x_{n-k} - x_{n-k-1}\| < +\infty. \qquad (6.33)$$

Let $\{x_n\}$ *be the sequence generated by the multi-step inertial Davis–Yin splitting method (6.32). Suppose that* $\inf_{n \geq 0} \lambda_n > 0$ *and* x^* *is the weak limit of* $\{x_n\}$. *Then the sequence* $\{J_{\gamma A_1} \circ (2J_{\gamma A_2} - I - \gamma B \circ J_{\gamma A_2})(x_n)\}$ *weakly converges to* $J_{\gamma A_2}(x^*) \in$ *zer*$(A_1 + A_2 + B)$.

Remark 6.12 Recently, with the tool of the partial smoothness, Poon and Liang [174] designed a framework to analyze the trajectory of the fixed point sequence generated by first-order methods, based on which they explained the reasons that some inertial schemes fail to provide acceleration. They also proposed a trajectory following adaptive acceleration algorithm, which is actually a multi-step inertial method. However, it is difficult to extend the results to general iterative methods, such as the Krasnosel'skiĭ–Mann iteration of the nonexpansive mappings.

Chapter 7
Relaxation Parameters of the Krasnosel'skiĭ–Mann Iteration

In this chapter, we present some ranges and several optimal choices of some relaxation parameter sequence $\{\lambda_n\}$ of the Krasnosel'skiĭ–Mann iteration (3.1) in the theory and actual practice. Some variants of the Krasnosel'skiĭ–Mann iteration are obtained based on the equivalence of the fixed point problem, the variational inequality problem, and nonlinear monotone equations.

The Krasnosel'skiĭ–Mann iteration is said to be:

(1) *Underrelaxed* if $\lambda_n \leq 1$.
(2) *Unrelaxed* if $\lambda_n \equiv 1$.
(3) *Overrelaxed* if $\lambda_n \geq 1$.

The relaxation parameter sequence $\{\lambda_n\}$ is of the paramount importance to the Krasnosel'skiĭ–Mann iteration. However, the results on how to select the relaxation parameter sequence $\{\lambda_n\}$ to speed up the convergence of the Krasnosel'skiĭ–Mann iteration are very limited, although it is shown that the convergence rate of the Krasnosel'skiĭ–Mann iteration depends heavily on the selection of the relaxation parameter sequence $\{\lambda_n\}$ (see, for example, [40]).

This problem was first investigated by Combettes [40] for the special nonexpansive mapping $P_D P_S$, i.e., $T = P_D P_S$ in the Krasnosel'skiĭ-Mann iteration (3.1), where D is a closed subspace, S is a closed convex subset of $\mathcal{H}$, and P_D and P_S are metric projection operators onto D and S, respectively.

Let p be a point in $\text{Fix}(P_D P_S)$. It was proved in [40] that the optimal relaxation parameter is given as follows:

$$\lambda_n^* = \frac{\langle x_n - P_D P_S(x_n), x_n - p \rangle}{\|x_n - P_D P_S(x_n)\|^2}. \tag{7.1}$$

It is observed that the optimal relaxation parameter sequence $\{\lambda_n^*\}$ in (7.1) depends on the fixed point p, which of course is not known. However, the range of the optimal relaxation parameter sequences can be narrowed down by using the formula (7.1), which is seen in the following lemma.

© The Author(s), under exclusive license to Springer Nature Switzerland AG 2022
Q.-L. Dong et al., *The Krasnosel'skiĭ-Mann Iterative Method*, SpringerBriefs in Optimization, https://doi.org/10.1007/978-3-030-91654-1_7

Lemma 7.1 ([40, Proposition 5]) *The optimal relaxations in the Krasnosel'skiĭ–Mann iteration:*

$$x_{n+1} = (1 - \lambda_n)x_n + \lambda_n P_D P_S(x_n) \ for \ each \ n \geq 0$$

are overrelaxation, i.e., $\lambda_n^* \geq 1$ for each $n \geq 0$.

It was claimed in [40] that, theoretically, the optimal relaxation parameter λ_n^* could be very large and definitely greater than 2.

Note that the optimal relaxation parameter at each iteration of the Krasnosel'skiĭ–Mann iteration does not systematically ensure faster convergence of the whole sequence of iterates to a fixed point. However, it was found that this is the case for the parallel projection method [40, 112], where it was found that the overrelaxations have an accelerating effect on the progression of the iterations toward a solution.

Since λ_n^* in (7.1) involves a fixed point and cannot be used in the practical computation, Combettes [40] used the so-called Armijo relaxation skill, which consists in reducing the relaxation parameter λ_n until the inequality:

$$\Phi(x_n) - \Phi(x_{n+1}) \geq \mu \lambda_n \|\nabla \Phi(x_n)\|^2 \qquad (7.2)$$

is satisfied, where $\mu \in (0, 1)$ is an arbitrary fixed constant and the function $\Phi :$ $\mathscr{H} \to [0, +\infty)$ is defined by

$$\Phi(x) = \frac{1}{2}\|x - P_D P_S(x)\|^2.$$

Note that the prerequisite of using (7.2) to give an approximate choice of λ_n is that $\Phi(x) = \frac{1}{2}\|x - P_D P_S(x)\|^2$ is differentiable. However, it is known that $\Phi(x) = \frac{1}{2}\|x - T(x)\|^2$ is not differentiable for a general nonexpansive or firmly nonexpansive mapping T.

Xu [198] also investigated this problem and his core idea is to regard the Krasnosel'skiĭ–Mann iteration as the line search method. More precisely, the algorithm (3.1) is rewritten in the form:

$$x_{n+1} = x_n - \lambda_n v_n \ for \ each \ n \geq 0,$$

where $v_n = x_n - T(x_n)$ is the search direction. He presented two choices of the relaxation parameter λ_n by solving optimization problems as follows: for each $n \geq 0$,

$$\lambda_n = \underset{0 \leq \lambda \leq c}{\operatorname{argmin}} \{\|x_n(\lambda) - T(x_n)(\lambda)\|^2 - \beta \lambda^2 \|x_n - T(x_n)\|\} \qquad (7.3)$$

and

$$\lambda_n = \operatorname*{argmin}_{0 \le \lambda \le 1}\{\|x_n(\lambda) - T(x_n)(\lambda)\|^2 - \beta\lambda(1-\lambda)\|x_n - T(x_n)\|\}, \qquad (7.4)$$

where $x_n(\lambda) := x_n - \lambda(x_n - T(x_n))$, $c \in (0,1)$ and $\beta > 0$ is a parameter.

In fact it is difficult to exactly solve the minimization problems (7.3) and (7.4) when T is a general nonlinear mapping. Furthermore, the numerical algorithms, such as the golden section search, may have a large amount of computation time when T is complex.

Recently, Iutzeler and Hendrickx [114] proposed the online relaxation Krasnosel'skiĭ–Mann iteration for an averaged mapping, which automatically tunes the relaxation parameters $\{\lambda_n\}$. Let $T : \mathbb{R}^N \to \mathbb{R}^N$ be a σ-averaged mapping.

Algorithm 3 The online relaxation Krasnosel'skiĭ–Mann iteration

---
Initialization: $\epsilon \in (0, 2\min(\sigma, 1-\sigma)]$, x_0, $x_1 = T(x_0)$, $\lambda_0 = \lambda_1 = 1$.
Compute

$$\lambda_{n+1} = \frac{(2-\epsilon)\lambda_n}{2\sigma\lambda_n + 1 - \frac{\lambda_{n-1}\|x_n - x_{n-1}\|}{\lambda_n\|x_{n-1} - x_{n-2}\|}} + \frac{\epsilon}{4\sigma},$$

$$(7.5)$$

$$x_{n+1} = (1 - \lambda_{n+1})x_n + \lambda_{n+1}T(x_n), \quad \text{for each } n \ge 1.$$

The convergence of the online relaxation Krasnosel'skiĭ–Mann iteration was established in following theorem.

Theorem 7.1 ([114, Theorem 4.1]) *Let $T : \mathbb{R}^N \to \mathbb{R}^N$ be a σ-averaged operator such that $\text{Fix}(T) \ne \emptyset$ for some $\sigma \in (0,1)$. Then the sequence $\{x_n\}$ generated by the online relaxation Krasnosel'skiĭ–Mann iteration converges to a point in $\text{Fix}(T)$.*

Giselsson et al. [86] proposed the following line search method for approximating relaxation parameters of Krasnosel'skiĭ–Mann iteration: for each $n \ge 0$,

$$\begin{cases} y_n = (1 - \bar{\lambda})x_n + \bar{\lambda}T(x_n) \\ x_{n+1} = (1 - \lambda_n)y_n + \lambda_n T(y_n), \end{cases} \qquad (7.6)$$

where $0 < \bar{\lambda} \le \lambda_{\max}$ is fixed and $\lambda_n \in [\bar{\lambda}, \lambda_{\max}]$ satisfies

$$\|x_{n+1} - T(x_{n+1})\| \le (1-\epsilon)\|y_n - T(y_n)\| \quad \text{for each } n \ge 0, \qquad (7.7)$$

where $\epsilon \in (0,1)$.

The algorithm (7.6) is in fact the two-step Krasnosel'skiĭ–Mann iteration, where the relaxation parameter in first step is fixed and that in second step is given via the linear search method (7.7).

Theorem 7.2 ([86, Theorem 2]) *Let $\mathscr{H}$ be a finite-dimensional real Hilbert space and $T : \mathscr{H} \to \mathscr{H}$ be a nonexpansive mapping with $\mathrm{Fix}(T) \neq \emptyset$. Suppose that $\bar{\lambda} \in (0, 1)$. Then the iteration (7.6) satisfies $\|x_n - T(x_n)\| \to 0$ and $\|x_{n+1} - x_n\| \to 0$ as $n \to \infty$.*

The fixed point residual $\|x_n - T(x_n)\|$ must be evaluated to carry out the line search test (7.7). In the general case, this requires us to evaluate the operator T, which has the same cost as a full iteration of the algorithm. Therefore, in the general case, it may be too expensive to evaluate many (or even just more than one) candidate relaxation parameters λ_n compared to the savings in iterations due to the line search.

A special case that $T = T_2 T_1$ is considered in [86], where the cost of evaluating T_2 and vector–vector operations are negligible (or at least, dominated by the cost of evaluating T_1). If T_1 is affine, the computational cost of the Krasnosel'skiĭ–Mann iteration with line search is the same as the basic iteration without the line search. Several line search variations, such as the linear search activation, are also presented in [86].

Recently, Themelis and Patrinos [190] proposed the SuperMann iteration by generalizing the classical Krasnosel'skiĭ–Mann iteration. Let $T : \mathscr{H} \to \mathscr{H}$ be α-averaged and $R = \mathrm{Id} - T$.

Algorithm 4 The SuperMann iteration

Require: $x_0 \in \mathscr{H}, c_0, c_1, q \in [0, 1), \beta, \sigma \in (0, 1), \lambda \in (0, \frac{1}{\alpha})$.
Initialize: $\eta_0 = r_{\text{safe}} = \|R(x_0)\|, n = 0$.
Step 1: If $R(x_n) = 0$, then stop.
Step 2: Choose an update direction $d_n \in \mathscr{H}$.
Step 3: (N_0) If $\|R(x_n)\| \leq c_0 \eta_n$, then set $\eta_{n+1} = \|R(x_n)\|$, proceed with a *blind update*

$$x_{n+1} = w_n := x_n + d_n$$

   and go to Step 6.
Step 4: Set $\eta_{n+1} = \eta_n$ and $\tau_n = 1$.
Step 5: Let $w_n = x_n + \tau_n d_n$.
   5(a) (N_1) If the *safe condition* $\|R(x_n)\| \leq r_{\text{safe}}$ holds and w_n is *educated*:

$$\|R(w_n)\| \leq c_1 \|R(x_n)\|.$$

   Then set $x_{n+1} = w_n$, update $r_{\text{safe}} = \|R(w_n)\| + q^n$, and go to Step 6.
   5(b) (N_2) If $\rho_n := \|R(w_n)\|^2 - 2\alpha \langle R(w_n), w_n - x_n \rangle \geq \sigma \|R(w_n)\| \|R(x_n)\|$.
   Then set

$$x_{n+1} = x_n - \lambda \frac{\rho_n}{\|R(w_n)\|^2} R w_n.$$

   Otherwise, set $\tau_n \leftarrow \beta \tau_n$ and go to Step 5.
Step 6: Set $n \leftarrow n + 1$ and go to Step 1.

Note that the directions d_n in the SuperMann iteration are superlinear (see Theorems VI.4 and VI.8 in [190]) and a natural choice of it is $-R(x_n)$. There are three types of updates: blind updates, educated updates, and safeguard updates in the SuperMann iteration. Although the SuperMann iteration was shown to converge superlinearly in [190], the convergence conditions are very restrictive, which prevents its applications in the actual computation.

7.1 The Approximate Optimal Relaxation Sequence

In this section, we discuss the optimal relaxation parameters introduced in [94] for the Krasnosel'skiĭ–Mann iteration of a nonexpansive mapping T.

Suppose that the n-th iterate x_n has been constructed and $p \in \text{Fix}(T)$ is taken arbitrarily. He et al. [94] used the same idea as that in [40] to obtain the optimal relaxation parameter λ_n by minimizing $\|x_{n+1} - p\|^2$, that is, guaranteeing that x_{n+1} is as close as possible to the fixed point p.

By using (2.4), it follows that

$$
\begin{aligned}
&\|x_{n+1} - p\|^2 \\
&= \|(1 - \lambda_n)(x_n - p) + \lambda_n(T(x_n) - p)\|^2 \\
&= (1 - \lambda_n)\|x_n - p\|^2 + \lambda_n\|T(x_n) - p\|^2 - \lambda_n(1 - \lambda_n)\|x_n - T(x_n)\|^2 \quad (7.8) \\
&= \lambda_n^2\|x_n - T(x_n)\|^2 - \lambda_n(\|x_n - p\|^2 - \|T(x_n) - p\|^2 + \|x_n - T(x_n)\|^2) \\
&\quad + \|x_n - p\|^2.
\end{aligned}
$$

This quadratic form in λ_n is minimized for

$$
\widehat{\lambda}_{p,n} = \frac{1}{2} + \frac{\|x_n - p\|^2 - \|T(x_n) - p\|^2}{2\|x_n - T(x_n)\|^2}, \tag{7.9}
$$

which can be seen the n-th optimal relaxation parameter of the Krasnosel'skiĭ–Mann iteration corresponding to the fixed point p since it involves p. It is easy to verify that $\widehat{\lambda}_{p,n}$ in (7.9) is the same as λ_n^* given in (7.1) for $T = P_D P_S$.

Next, we introduce optimal relaxation parameters that do not depend on the fixed point p for a closed affine set and a closed convex set, respectively.

Definition 7.1 Let C be a closed affine set of $\mathcal{H}$ and $T : C \to C$ be a nonexpansive mapping. For each $n \geq 0$, suppose the n-th iterate x_n has been constructed. Then the parameter defined by

$$
\widehat{\lambda}_n = \inf_{p \in \text{Fix}(T)} \widehat{\lambda}_{p,n} \tag{7.10}
$$

is said to be the *n-th optimal relaxation parameter* of the Krasnosel'skiĭ–Mann iteration (3.1), where $\widehat{\lambda}_{p,n}$ is given by (7.9).

Let C be a general closed convex set and $T : C \to C$ be a nonexpansive (or σ-averaged) mapping. Since $\widehat{\lambda}_{p,n}$ often overpasses $[0, 1]$ (or $[0, 1/\sigma]$), to guarantee $\{x_n\} \subset C$, some restrictions are made in the below definition.

Definition 7.2 Let C be a closed convex set and $T : C \to C$ be a nonexpansive mapping. For each $n \geq 0$, suppose the n-th iterate x_n has been constructed. Then the parameter $\widehat{\lambda}_n$ defined by

$$\widehat{\lambda}_n = \begin{cases} \min\{\inf_{p \in \mathrm{Fix}(T)} \widehat{\lambda}_{p,n}, 1\}, & \text{if } T \text{ is nonexpansive}, \\ \min\{\inf_{p \in \mathrm{Fix}(T)} \widehat{\lambda}_{p,n}, \frac{1}{\sigma}\}, & \text{if } T \text{ is } \sigma\text{-averaged}, \end{cases} \qquad (7.11)$$

is said to be the *n-th optimal relaxation parameter* of the Krasnosel'skiĭ–Mann iteration (3.1), where $\widehat{\lambda}_{p,n}$ is given by (7.9) and σ is some positive constant in $(0, 1)$.

The optimal relaxation parameter sequence $\{\widehat{\lambda}_n\}$ given in (7.10) or (7.11) cannot be used in the actual computing, since it involves fixed points, and thus we are not able to get it in advance. However, by using the property of the operator T, one may obtain the strategy and manner for the relaxation parameters' selection of the Krasnosel'skiĭ–Mann iteration.

He et al. [94] presented some fundamental properties of the optimal relaxation parameter sequence $\{\widehat{\lambda}_n\}$.

Lemma 7.2 ([94, Lemma 3.1]) *Let C be a closed convex set, $T : C \to C$ be a mapping, and $\{\widehat{\lambda}_n\}$ be the optimal relaxation parameter sequence given by (7.10) or (7.11) for the Krasnosel'skiĭ–Mann iteration (3.1). Then we have the following:*

(1) *If T is a nonexpansive mapping, then $\widehat{\lambda}_n \geq \frac{1}{2}$ holds.*
(2) *If T is σ-averaged for some constant $\sigma \in (0, 1]$, then $\widehat{\lambda}_n \geq \frac{1}{2\sigma}$ for each $n \geq 0$.*
(3) *If T is a quasi-isometric mapping, i.e., $\|T(x) - p\| = \|x - p\|$ holds for all $x \in C$ and $p \in Fix(T)$, then, for each $n \geq 0$, $\widehat{\lambda}_n = \frac{1}{2}$.*
(4) *Denote by $ran(\mathrm{Id} - T)$ the range of $\mathrm{Id} - T : C \to H$. If $\mathrm{Id} - T$ is injective and its inverse operator $(\mathrm{Id} - T)^{-1} : ran(\mathrm{Id} - T) \to C$ is L-Lipschitz continuous, then*

$$\widehat{\lambda}_n \leq L \ \text{ for each } n \geq 0.$$

From Lemma 7.2 (2), it follows that the lower bound of $\{\widehat{\lambda}_n\}$ becomes bigger as the averaged constant σ decreases. So, we have to determine the minimum averaged constant of an averaged operator to get the maximum lower bound of $\{\widehat{\lambda}_n\}$.

By substituting p in $\|x_n - p\|^2$ and $\|T(x_n) - p\|^2$ in (7.9) with $T(x_n)$ and $T^2(x_n)$, respectively, He et al. [94] presented the following approximation to the optimal relaxation parameter sequence $\widehat{\lambda}_n$ in (7.9):

$$\lambda_n = \frac{1}{2} + \frac{\|x_n - T(x_n)\|^2 - \|T(x_n) - T^2(x_n)\|^2}{2\|(x_n - T(x_n)) - (T(x_n) - T^2(x_n))\|^2} \quad \text{for each } n \geq 0. \qquad (7.12)$$

The numerical results in [94] show the advantage of the approximate relaxation parameters (7.12) comparing with the fixed relaxation parameter and that given in (7.5).

The following theorem shows the convergence of the Krasnosel'skiĭ–Mann iteration (3.1) with the approximate optimal parameter sequence $\{\lambda_n\}$ given by (7.12) under some conditions.

Theorem 7.3 *Let C be a closed affine set, and let $T : C \to C$ be a nonexpansive mapping with $Fix(T) \neq \emptyset$. If the optimal parameter $\widehat{\lambda}_n$ given by (7.10) and the approximate optimal parameter λ_n given by (7.12) satisfy*

$$\lambda_n \leq (2 - \varepsilon)\widehat{\lambda}_n$$

for some $\varepsilon \in (0, 1)$ for each $n \in \mathbb{N}$, then the sequence $\{x_n\}$ generated by the Krasnosel'skiĭ–Mann iteration (3.1) with the approximate optimal parameter sequence $\{\lambda_n\}$ given by (7.12) converges weakly to a fixed point of T.

The condition $\lambda_n \leq (2 - \varepsilon)\widehat{\lambda}_n$ in Theorem 7.3 can be gotten rid when $T : \mathbb{R}^m \to \mathbb{R}^m$ is an affine linear operator.

Theorem 7.4 *Assume that $T : \mathbb{R}^m \to \mathbb{R}^m$ is an affine linear operator defined by $T(x) = Bx + d$ for all $x \in \mathbb{R}^m$, where B is an $m \times m$ real symmetric matrix with the eigenvalues $\lambda_i \in (-1, 1]$ for each $i = 1, \cdots, m$ and $d \in \mathbb{R}^m$ is a given vector. Let $\{x_n\}$ be the sequence generated by the Krasnosel'skiĭ–Mann iteration (3.1) with the approximate optimal parameter sequence $\{\lambda_n\}$ given by (7.12). Then we have the following:*

(1) *The sequence $\{x_n\}$ converges to a fixed point of T.*
(2) *There holds the convergence rate estimate:*

$$\|x_n - T(x_n)\| \leq \rho^n \|x_0 - T(x_0)\| \quad \text{for each } n \geq 0, \qquad (7.13)$$

where

$$\rho = \sqrt{1 - \left(\frac{1 - \bar{\lambda}}{1 - \underline{\lambda}}\right)^2}, \quad \bar{\lambda} = \max_{i \in J} \lambda_i, \quad \underline{\lambda} = \min_{i \in J} \lambda_i,$$

$$J = \{i : i = 1, \cdots, m, \lambda_i \neq 1\}.$$

Let C be a general closed convex set. In order to obtain the convergence of the Krasnosel'skiĭ–Mann iteration (3.1) with the approximate optimal parameter sequence $\{\lambda_n\}$ given by (7.12), He et al. [94] imposed the upper bound on the sequences of the relaxation parameters λ_n in (7.12).

Theorem 7.5 *Let C be a closed convex set of a Hilbert space $\mathcal{H}$ and $T : C \to C$ be a nonexpansive mapping with $Fix(T) \neq \emptyset$. Let $\{\varepsilon_n\} \subset (0, 1)$ be the sequence such that*

$$\sum_{n=0}^{\infty} \varepsilon_n(1 - \varepsilon_n) = \infty.$$

Then the sequence $\{x_n\}$ generated by the Krasnosel'skiĭ–Mann iteration (3.1) converges weakly to a fixed point of T, where the relaxation parameter sequence $\{\lambda_n\}$ is selected by the formula:

$$\lambda_n = \min\left\{\frac{1}{2} + \frac{\|x_n - T(x_n)\|^2 - \|T(x_n) - T^2(x_n)\|^2}{2\|(x_n - T(x_n)) - (T(x_n) - T^2(x_n))\|^2}, \; 1 - \varepsilon_n\right\}. \qquad (7.14)$$

Theorem 7.6 *Let C be a closed convex set of a Hilbert space $\mathcal{H}$ and $T : C \to C$ be a σ-averaged (nonexpansive) mapping with $Fix(T) \neq \emptyset$ for some positive constant $\sigma \in (0, 1)$. Let $\{\varepsilon_n\} \subset (0, 1)$ be the sequence such that*

$$\sum_{n=0}^{\infty} \varepsilon_n(1 - \varepsilon_n) = \infty.$$

Then the sequence $\{x_n\}$ generated by the Krasnosel'skiĭ–Mann iteration (3.1) converges weakly to a fixed point of T, where the relaxation parameter sequence $\{\lambda_n\}$ is selected by the formula:

$$\lambda_n = \min\left\{\frac{1}{2} + \frac{\|x_n - T(x_n)\|^2 - \|T(x_n) - T^2(x_n)\|^2}{2\|(x_n - T(x_n)) - (T(x_n) - T^2(x_n))\|^2}, \; \frac{1}{\sigma}(1 - \varepsilon_n)\right\}. \qquad (7.15)$$

The numerical examples in [94] illustrate that the upper bound imposed on the relaxation parameter (7.12) generally affects the convergence speed of the Krasnosel'skiĭ–Mann iteration.

To this end, He et al. [94] proposed the following open problem.

Open Question *Whether does the Krasnosel'skiĭ–Mann iteration (3.1) with the approximate optimal parameter sequence $\{\lambda_n\}$ given by (7.12) converge?*

A shortcoming of the relaxation parameters given in (7.12) is that it involves the operator T, so the computational cost may be large when T is complicated. However, from the numerical examples in [94], it is observed that the relaxation parameter sequence in (7.12) varies little or oscillates in a small range after some iterations, so one can use the averaged value of the relaxation parameter sequence at first some iterations to substitute it.

7.2   Some Variants of the Krasnosel'skiĭ–Mann Iteration Based on the Projection Methods of Variational Inequality Problems

In this section, we mainly discuss the relation of the fixed point problem (1.1) and the variational inequality problems, based on which the projection methods for the variational inequality problems can be extended to obtain the variants of the Krasnosel'skiĭ–Mann iteration.

The fixed point problem has a very close relation with the classical *variational inequality problem* introduced by Stampacchia [185] as follows:

$$\text{Find } x^* \in C \text{ such that } \langle F(x^*), y - x^* \rangle \geq 0 \text{ for all } y \in C, \tag{7.16}$$

where C is a nonempty closed and convex subset of a Hilbert space $\mathcal{H}$ and $F : \mathcal{H} \to \mathcal{H}$ is an operator. Denote by $SOL(C, F)$ the solution of variational inequality problem (7.16). Indeed, by Lemma 2.3, the problem (7.16) equals to the following:

$$x^* = P_C(x^* - F(x^*)). \tag{7.17}$$

Let $T : C \to C$ be a mapping and

$$F := \text{Id} - T.$$

Then, from (7.17), it follows that

$$x^* = T(x^*). \tag{7.18}$$

Therefore, we get the conclusion that the fixed point problem (7.18) with $T : C \to C$ is equivalent to the variational inequality problem (7.16) with $F = \text{Id} - T$ (see, for example, [121, 132] for more discussion).

Let $T : C \to C$ be a nonexpansive mapping. Then some properties of $F := \text{Id} - T$ are given in the following lemma.

Lemma 7.3 *Let* $T : C \to C$ *be a nonexpansive mapping. Then* F *is monotone and 2-Lipschitz continuous.*

Proof For all $x, y \in \mathcal{H}$, due to the nonexpansivity of T, we have

$$\langle F(x) - F(y), x - y \rangle = \|x - y\|^2 - \langle T(x) - T(y), x - y \rangle$$
$$\geq \|x - y\| \left(\|x - y\| - \|T(x) - T(y)\| \right)$$
$$\geq 0,$$

which implies that F is monotone. Using the nonexpansivity of T again,

$$\|F(x) - F(y)\| = \|(x - y) - (T(x) - T(y))\|$$
$$\leq \|x - y\| + \|T(x) - T(y)\|$$
$$\leq 2\|x - y\|.$$

So, F is 2-Lipschitz continuous. This completes the proof.

The following result shows the relation of the set of fixed point of T and the solution set of the corresponding variational inequality problem.

Lemma 7.4 *Let $T : C \to C$ be a nonexpansive mapping with $\mathrm{Fix}(T) \neq \emptyset$ and $F = \mathrm{Id} - T$. Then $\mathrm{Fix}(T) = SOL(C, F)$.*

Proof It is easy to verify that $\mathrm{Fix}(T) \subseteq SOL(C, F)$.
  Next, we show

$$SOL(C, F) \subseteq \mathrm{Fix}(T).$$

In fact, take arbitrarily $x^* \in SOL(C, F) \subseteq C$. For all $\alpha \in (0, 1)$, from (7.16), it follows that

$$\langle x^* - (x^* - \alpha F(x^*)), x - x^* \rangle \geq 0 \ \text{for all} \ x \in C.$$

Using Lemma 2.7, we have

$$x^* = P_C(x^* - \alpha F(x^*)). \tag{7.19}$$

By the convexity of C and $\alpha \in (0, 1)$,

$$x^* - \alpha F(x^*) = (1 - \alpha)x^* + \alpha T(x^*) \in C. \tag{7.20}$$

Therefore, (7.19) yields $x^* = x^* - \alpha F(x^*)$, i.e., $F(x^*) = 0$. Thus we get $x^* = T(x^*)$ and hence $x^* \in \mathrm{Fix}(T)$. So, $SOL(C, F) \subseteq \mathrm{Fix}(T)$. This completes the proof.

There are a great deal of projection methods for the variational inequality problem with monotone and Lipschitz continuous operator, the most famous of which is the extragradient method introduced by Korpelevich [117]. The extragradient method has received a great deal of attention from many authors, who improved it in various ways (see, for example, [30, 59, 71, 93, 113, 116, 183, 192] and references therein). Most of these algorithms involve two operators or two projections each iteration.

Recently, some algorithms were introduced, which just involve one projection and one operator each iteration [139, 201, 202]. By using the equivalent relation of the fixed point problem and the variational inequality problem, all the methods for the variational inequality problem can be extended to the fixed point problem. Here we give three formulas for solving the fixed point problem based on the recently advanced three typical iterative algorithms of the variational inequality problem.

First, we present an iterative algorithm based on Algorithm A in [201]. Let $T : \mathscr{H} \to \mathscr{H}$ be a nonexpansive mapping.

Algorithm 5

Step 0: Take $\delta \in (1, +\infty)$ and choose $x_0 = y_0 \in \mathscr{H}$, $\lambda_0 > 0$, $\sigma \in (0, \frac{\sqrt{2}-1}{\delta})$.

Step 1: Given the current iterates x_n and y_n for each $n \geq 0$, compute

$$x_{n+1} = x_n + \lambda_n(T(y_n) - y_n). \qquad (7.21)$$

Step 2: If $x_n = y_n = x_{n+1}$, then stop: x_n is a fixed point of T. Otherwise, compute

$$y_{n+1} = x_{n+1} + \delta(x_{n+1} - x_n), \qquad (7.22)$$

and

$$\lambda_{n+1} = \begin{cases} \min\left\{ \dfrac{\sigma\|y_{n+1} - y_n\|}{\|y_{n+1} - y_n - Ty_{n+1} + Ty_n\|}, \lambda_n \right\}, \\ \qquad\qquad \text{if } y_{n+1} - y_n - Ty_{n+1} + Ty_n \neq 0, \\ \lambda_n, \qquad\qquad \text{otherwise.} \end{cases} \qquad (7.23)$$

Set $n := n + 1$ and return to Step 1.

Note that T maps $\mathscr{H}$ into itself, so there does not need the projection in Algorithm 5, which appears Algorithm A in [201].

A simple calculation of (7.21) and (7.22) implies that, for each $n \geq 0$,

$$\begin{aligned} x_{n+1} &= (1 - \lambda_n)\left[x_n - \delta\frac{\lambda_n}{1 - \lambda_n}(x_n - x_{n-1}) \right] + \lambda_n T(y_n) \\ &= (1 - \lambda_n)z_n + \lambda_n T(y_n), \end{aligned}$$

where

$$z_n = x_n - \delta\frac{\lambda_n}{1 - \lambda_n}(x_n - x_{n-1}) \quad \text{for each } n \geq 0.$$

Remark 7.1 It is easy to see that Algorithm 5 is a special case of the general inertial Krasnosel'skiĭ–Mann iteration (5.6). There are two illustrations for the relaxation parameter and inertial parameter:

(1) The relaxation parameter λ_n is given by the adaptive way (7.23), and the sequence $\{\lambda_n\}$ is monotonically decreasing with the lower bound $\min\{2\sigma, \lambda_0\}$ (see [201, Remark 2.1] for details).

(2) The inertial parameter $-\delta\frac{\lambda_n}{1-\lambda_n}$ in z_n may be negative and involves the relaxation parameter λ_n, and the inertial parameter δ in y_n is greater than 1.

The convergence result of Algorithm 5 can be obtained by combining [201, Theorem 3.1] and Lemmas 7.3 and 7.4.

Theorem 7.7 *Let $T : \mathscr{H} \to \mathscr{H}$ be a nonexpansive mapping with Fix(T)$\neq \emptyset$. Then the sequence $\{x_n\}$ generated by Algorithm 5 converges weakly to a point in Fix(T).*

If we make further assumption on the operator T as follows:

$$\langle T(x) - T(y), x - y \rangle \leq (1 - M)\|x - y\|^2 \quad \text{for all } x, y \in \mathscr{H} \tag{7.24}$$

for some $M > 0$, then Algorithm 5 has the R-linear rate of the convergence.

Theorem 7.8 *Let $T : \mathscr{H} \to \mathscr{H}$ be a nonexpansive mapping satisfying (7.24). Then the sequence $\{x_n\}$ generated by Algorithm 5 converges to the unique fixed point of T at least R-linearly.*

Next, we present an iterative algorithm, i.e., Algorithm 6, by using Algorithm 4.1 in [202].

Algorithm 6

Step 0: Choose $\lambda_0 > 0$, $x_0, y_0 \in \mathscr{H}$, $\mu, \sigma \in (0, 1)$, $\theta \in (0, 1]$, and

$$\delta \in \left(\frac{\sqrt{1 + 4(\frac{\sigma}{2 - \theta} + 1 - \sigma)} - 1}{2}, 1 \right).$$

Step 1: Given the current iterates x_n and y_n for each $n \geq 1$, compute

$$\begin{cases} y_{n+1} = (1 - \delta)x_n + \delta y_n, \\ x_{n+1} = y_{n+1} - \lambda_n x_n + \lambda_n T(x_n). \end{cases} \tag{7.25}$$

If $y_{n+1} = x_n = x_{n+1}$ (or $T(x_n) = x_n$), then stop: x_n is a fixed point of T. Otherwise, go to Step 2.

Step 2: Compute

$$\lambda_{n+1} = \begin{cases} \min \left\{ \dfrac{\sigma \mu \theta (\|x_n - x_{n-1}\|^2 + \|x_{n+1} - x_n\|^2)}{4\delta \langle x_{n-1} - x_n - T(x_{n-1}) + T(x_n), x_{n+1} - x_n \rangle}, \lambda_n \right\}, \\ \qquad \text{if } \langle x_{n-1} - x_n - T(x_{n-1}) + T(x_n), x_{n+1} - x_n \rangle \geq 0, \\ \lambda_n, \qquad \text{otherwise.} \end{cases} \tag{7.26}$$

Set $n := n + 1$ and return to Step 1.

Set $y_0 = x_0$ in Algorithm 6. By Remark 6.8 (1), we get

$$y_{n+1} = x_n - \sum_{k=1}^{n} \delta^{n-k+1}(x_k - x_{k-1}).$$

By putting it with the second formula of (7.25), we get

$$\begin{cases} z_n = x_n - \dfrac{1}{1 - \lambda_n} \sum_{k=1}^{n} \delta^{n-k+1}(x_k - x_{k-1}), \\ x_{n+1} = (1 - \lambda_n)z_n + \lambda_n T(x_n). \end{cases}$$

Therefore, it is easy to see that Algorithm 6 with $y_0 = x_0$ is a special case of the multi-step inertial Krasnosel'skiĭ–Mann iteration.

Note that the schemes (7.25) and (6.22) are different since the relaxation parameters and inertial parameters are not the same.

Similar to Algorithm 5, the relaxation parameter λ_n is given by the adaptive way (7.26) and has a lower bound $\min\{\frac{\sigma\mu\theta}{4\delta}, \lambda_1\}$. The inertial parameters may be negative and involve the relaxation parameters.

The convergence result of Algorithm 6 can be obtained by combining [202, Theorem 4.1] and Lemmas 7.3 and 7.4:

Theorem 7.9 *Let $\mathscr{H}$ be a finite-dimensional real Hilbert space and $T : \mathscr{H} \to \mathscr{H}$ be a nonexpansive mapping with $Fix(T) \neq \emptyset$. Then the sequences $\{x_n\}$ and $\{y_n\}$ generated by Algorithm 6 converge to the same point in $Fix(T)$.*

Now, we propose an iterative algorithm, i.e., Algorithm 7, by using the golden ratio algorithms for the variational inequality problem in [137].

Let $T : \mathscr{H} \to \mathscr{H}$ be a nonexpansive mapping. Let $\varphi = \frac{\sqrt{5}+1}{2}$ be the golden ratio, that is, $\varphi^2 = 1 + \varphi$.

Algorithm 7

Input: Choose $x_0, x_1 \in \mathscr{H}$, $\lambda_0 > 0$, $\phi \in (1, \varphi]$, and $\bar{\lambda} > 0$. Set $y_0 = x_1$, $\theta_0 = 1$, and
$\rho = \frac{1}{\phi} + \frac{1}{\phi^2}$.
For each $k \geq 1$, do
Step 1: Find the stepsize:

$$\lambda_n = \left\{ \rho\lambda_{n-1}, \frac{\phi\theta_{k-1}}{4\lambda_{k-1}} \frac{x_n - x_{n-1}}{\|x_n - x_{n-1} - T(x_n) + T(x_{n-1})\|^2}, \bar{\lambda} \right\}.$$

Step 2: Compute the next iterates:

$$\begin{cases} y_{n+1} = \left(1 - \dfrac{1}{\phi}\right) x_n + \dfrac{1}{\phi} y_n \\[2mm] x_{n+1} = y_{n+1} - \lambda_n x_n + \lambda_n T(x_n). \end{cases} \tag{7.27}$$

Step 3: Update: $\theta_n = \frac{\lambda_n}{\lambda_{n-1}}\phi$.

Comparing (7.25) with (7.27), it follows that the iterative schemes in Algorithms 6 and 7 are the same, while the relaxation parameter λ_n is different.

Notice that for $\phi \leq \varphi$ one has $\rho \geq 1$ and hence λ_n can be larger than λ_{n-1}. This may be the biggest difference of Algorithm 7, and Algorithms 5 and 6. The relaxation parameters in the latter two are nonincreasing, which is rather restrictive.

By combining Lemmas 7.3 and 7.4 and [137, Theorem 4], we obtain the convergence of Algorithm 7.

Theorem 7.10 *Let $\mathscr{H}$ be a finite-dimensional real Hilbert space and $T : \mathscr{H} \to \mathscr{H}$ be a nonexpansive mapping with $Fix(T) \neq \emptyset$. Then the sequence $\{x_n\}$ generated by Algorithm 7 converges to a point in $Fix(T)$.*

Until to now, there are some research on the range and choice of the inertial parameters and relaxation parameters of the (multi-step) inertial Krasnosel'skiĭ–Mann iteration. However there are seldom results on the optimal choices of the inertial parameters and relaxation parameters; therefore, here we propose the following problem.

Open Problem *How to choose inertial parameters and relaxation parameters of the (multi-step) inertial Krasnosel'skiĭ–Mann iteration to accelerate the convergence?*

7.3　The Residual Algorithm

In this section, we introduce a line search method for the relaxation parameters of the Krasnosel'skiĭ–Mann iteration based on the equivalence of the fixed point problem (1.1) and the system of nonlinear monotone equations.

The system of nonlinear monotone equations can be mathematically formulated as

$$\text{Find } x \in \mathscr{H} \text{ such that } F(x) = 0, \tag{7.28}$$

where $F : \mathscr{H} \to \mathscr{H}$ is a nonlinear operator.

The fixed point problem (1.1) has a close relation with the system of nonlinear monotone equations (7.28). Let $F = \text{Id} - T$. Then the problem (7.28) becomes the fixed point problem (1.1). By using this relation, the iterative algorithms for the system of nonlinear monotone equations (7.28) can be extended to solve the fixed point problem (1.1).

By extending the algorithm in [122] to the fixed point problem, La Cruz [121] introduced the following residual algorithm (Algorithm 8).

It is clear that $\mu_n \in (0, 1]$ for each $n \geq 0$. The scalar ρ_n is called *spectral coefficient*, which is closely related to the Barzilai–Borwein choice of the stepsize ([13, 54]) with $\rho_{\max} \gg 1$ sufficiently large. From the theoretical point of view, this choice of ρ_n ensures the asymptotic regularity

$$\lim_{n \to \infty} \|x_n - T(x_n)\| = 0.$$

It is easy to see that Algorithm 8 is exactly the Krasnosel'skiĭ–Mann iteration with the relaxation parameter $\lambda_n = \rho_n \mu_n$. Note that μ_n is obtained via the line search, which is an auxiliary iterative procedure that runs in each iteration of the algorithm until the criterion (7.29) is satisfied. In fact, the line search can be quite costly in general as it requires computing additional values of T.

To present the convergence of Algorithm 8, La Cruz [121] made the following assumption.

Algorithm 8

Step 0: Choose $x_0 \in \mathscr{H}$, $1 \ll \rho_{max} < \infty$, $\rho_0 \in (0, \rho_{max}]$, $\gamma, \sigma \in (0, 1)$, and a positive sequence $\{\eta_n\}$ such that

$$\sum_{n=0}^{+\infty} \eta_n = \eta < +\infty.$$

Let $n := 0$ and $y_0 = x_0 - T(x_0)$.

Step 1: If $\|x_n - T(x_n)\| = 0$, stop.

Step 2: Compute x_{n+1}, y_{n+1}, and ρ_{n+1} as follows: Find m_n as the smallest nonnegative integer number m such that

$$\|x_n - \sigma^m \rho_n y_n - T(x_n - \sigma^m \rho_n y_n)\|^2 \leq (1 - \gamma\sigma^{2m})\|y_n\|^2 + \eta_n. \tag{7.29}$$

Set $\mu_n = \sigma^{m_n}$ and

$$x_{n+1} = (1 - \lambda_n)x_n + \lambda_n T(x_n), \tag{7.30}$$

where $\lambda_n = \rho_n \mu_n$ and

$$\rho_{n+1} = \begin{cases} \dfrac{\lambda_n \|y_n\|^2}{\|y_n\|^2 - \langle y_{n+1}, y_n \rangle}, & \text{if } \langle y_{n+1}, y_n \rangle \leq \left(1 - \dfrac{\rho_n}{\rho_{max}}\right)\|y_n\|^2, \\ \rho_{max}, & \text{otherwise}, \end{cases}$$

and $y_{n+1} = x_{n+1} - T(x_{n+1})$.

Step 3: Set $n := n + 1$ and go to Step 1.

Assumption 3.1 Assume that the mapping $T : \mathscr{H} \to \mathscr{H}$ is a nonexpansive mapping with $\text{Fix}(T) \neq \emptyset$ and the set

$$\Omega = \{x \in \mathscr{H} : \|x - T(x)\|^2 \leq \|x_0 - T(x_0)\|^2 + \eta\}$$

is compact.

Assumption 3.1 seems restrictive and the author in [121] has not provided an example satisfying it.

Theorem 7.11 *If Assumption 3.1 holds, then the sequence $\{x_n\}$ generated by Algorithm 8 converges weakly to a point in $\text{Fix}(T)$.*

Chapter 8
Two Applications

This chapter discusses two applications of the Krasnosel'skiĭ–Mann iteration: asynchronous parallel coordinate updates methods and cyclic coordinate update algorithms. Their convergence analysis and convergence rate estimate are established under some conditions.

8.1 The Asynchronous Parallel Coordinate Updates Method

The asynchronous parallel method was firstly introduced by Chazan and Miranker [36] in 1969 and successfully applied in many fields, such as linear systems [8], nonlinear problems [9], differential equations [77], consensus problems [81], and optimization [131, 188].

Comparing the synchronous parallel iterative algorithms, the agents in asynchronous parallel iterative algorithms run continuously with little idling and the iterations are disordered where an agent may carry out an iteration without the newest information from other agents. Therefore, the asynchronous parallel iterative algorithms are easy to incorporate new agents and its system is more tolerant to computing faults and communication glitches.

The async-parallel iterative methods for the fixed point problem was proposed by Baudet [14], where the operator was assumed to be P-contraction. To solve the fixed point problem of nonexpansive mappings, Peng et al. [169] recently introduced an algorithmic framework for the asynchronous parallel coordinate updates methods (ARock for short), which can be seen an important application of the Krasnosel'skiĭ–Mann iteration.

Let $\mathscr{H}_1, \cdots, \mathscr{H}_m$ be Hilbert spaces and $\mathscr{H} := \mathscr{H}_1 \times \cdots \times \mathscr{H}_m$ be their Cartesian product.

For a nonexpansive operator $T : \mathscr{H} \to \mathscr{H}$, consider the fixed point problem:

© The Author(s), under exclusive license to Springer Nature Switzerland AG 2022
Q.-L. Dong et al., *The Krasnosel'skiĭ-Mann Iterative Method*, SpringerBriefs in Optimization, https://doi.org/10.1007/978-3-030-91654-1_8

$$\text{Find } x^* \in \mathscr{H} \text{ such that } T(x^*) = x^*. \tag{8.1}$$

Let $S = \text{Id} - T$, then finding a fixed point of T is equivalent to finding a zero of S, denoted by x^*, such that

$$0 = S(x^*).$$

The Krasnosel'skiĭ–Mann iteration (3.1) with $\lambda_n \equiv \lambda \in (0, 1)$ for each $n \geq 0$ can be equivalently written as follows:

$$x_{n+1} = x_n - \lambda S(x_n) \text{ for each } n \geq 0. \tag{8.2}$$

Let $(p_1, \cdots, p_m)$ be a distribution. Consider the n-th update applied to $x_{i_n} \in \mathscr{H}_{i_n}$, where $i_n \in \{1, \cdots, m\}$ is randomly selected according to the distribution $(p_1, \cdots, p_m)$. Each coordinate update in the ARock has the form:

$$x_{n+1} = x_n - \frac{\eta_n}{mp_{i_n}} S_{i_n}(\hat{x}_n) \text{ for each } n \geq 0, \tag{8.3}$$

where $\eta_n > 0$ is a scalar and $S_{i_n}(x) := (0, \cdots, 0, (S(x))_{i_n}, 0, \cdots, 0)$ and mp_{i_n} is used to normalize nonuniform selection probabilities. In the uniform case, namely, $p_i \equiv \frac{1}{m}$ for each $i \in \{1, 2, \cdots, m\}$, we have $mp_{i_n} \equiv 1$, which simplifies the update (8.3) to the following:

$$x_{n+1} = x_n - \eta_n S_{i_n}(\hat{x}_n) \text{ for each } n \geq 0. \tag{8.4}$$

Here the point $\hat{x}_n$ is what an agent reads from global memory to its local cache and to which S_{i_n} is applied, and x_n denotes the state of x in global memory just before the update (8.3) is applied. In the sync-parallel algorithm, we have $\hat{x}_n = x_n$, but, in the ARock, due to possible updates to x by other agents, $\hat{x}_n$ can be different from x_n. This is a key difference between the sync-parallel and async-parallel algorithms. The relation of $\hat{x}_n$ and x_n is as follows:

$$\hat{x}_n = x_n + \sum_{d \in J(n)} (x^d - x^{d+1}) \text{ for each } n \geq 1, \tag{8.5}$$

where $J(n) \subseteq \{n - 1, \cdots, n - \tau\}$ and $\tau \in \mathbb{N}^+$ denotes the maximum delay, i.e., the maximum number of other updates to x during the computation of (8.3).

A framework for the asynchronous parallel coordinate updates is given as follows:

Algorithm 9 ARock: A framework for the asynchronous parallel coordinate updates

---

Input: $x_0 \in \mathcal{H}$, $N > 0$, a distribution $(p_1, \cdots, p_m) > 0$ with $\sum_{i=1}^{m} p_i = 1$;
    global iteration counter $n \leftarrow 0$;
While $n < N$, *every agent asynchronously* **do**
    select $i_n \in \{1, \ldots, m\}$ with $\text{Prob}(i_n = i) = p_i$;
    perform an update to x_{i_n} according to (8.3);
    update the global counter $n \leftarrow n + 1$.

Note that, in Algorithm 9, a set of p agents, $p \geq 1$, solves the problem (8.1) by updating the coordinates $x_i \in \mathcal{H}_i$ for each $i = 1, \cdots, m$ in a random and asynchronous fashion. Whenever an agent finishes updating a coordinate, the global iteration counter n increases by one.

There are two assumptions on the update and delay for Algorithm 9:

(1) *Atomic coordinate update*: a coordinate is not further broken into smaller components during an update; they are all updated at once.
(2) *Bounded maximal delay* τ: during any update cycle of an agent, x in global memory is updated at most τ times by other agents.

For Algorithm 9, the computation requires the following:

(a) Random coordinate selection
(b) A finite maximal delay τ
(c) Bounded relaxation parameters

Assumption (a) is essential to both the analysis and the numerical performance of Algorithm 9. Assumption (b) is not essential and an infinite delay with a light tail is allowed (see [168]). Assumption (c) is standard.

Note that the update (8.3) is only computationally worthy if $S_i(x)$ is much cheaper to compute than $S(x)$, such as the corresponding operators in the system of linear equations, decentralized consensus optimization, and feasibility problem (see [169, Section 2] for details). Otherwise, it is more preferable to apply the full Krasnosel'skiĭ–Mann update (8.2).

To discuss the convergence of Algorithm 9, some notations are needed. $(\Omega, \mathcal{F}, P)$ denotes the underlying probability space, where Ω, $\mathcal{F}$, and P are the sample space, σ-algebra, and probability measure, respectively. The mapping $x : (\Omega, \mathcal{F}) \rightarrow (\mathcal{H}, \mathcal{B})$ is an $\mathcal{H}$-valued random variable, where $\mathcal{B}$ is the Borel σ-algebra. Let $\{x_n\}$ denote either a sequence of deterministic points in $\mathcal{H}$ or a sequence of $\mathcal{H}$-valued random variables, and let $x_i \in \mathcal{H}_i$ denote the ith coordinate of x. Let $\mathcal{X}_n := \sigma(x_0, \hat{x}_1, x_1, \cdots, \hat{x}_n, x_n)$ denote the smallest σ-algebra generated by $x_0, \hat{x}_1, x_1, \cdots, \hat{x}_n, x_n$. "Almost surely" is abbreviated as "a.s.," and the n product space of $\mathcal{H}$ is denoted by $\mathcal{H}^n$. In the product space, let

$$\mathbf{X}^* := \{(x^*, x^*, \cdots, x^*) : x^* \in \text{Fix}(T)\} \subseteq \mathcal{H}^{\tau+1}.$$

To obtain convergence results, we assume $p_{\min} := \min_i p_i > 0$ and

$$\text{Prob}(i_n = i \mid \mathscr{X}_n) = \text{Prob}(i_n = i) = p_i \quad \text{for each } i \geq 1 \text{ and } n \geq 0.$$

The following lemma develops an upper bound for the expected distance between x_{n+1} and any $x^* \in \text{Fix}T$:

Lemma 8.1 *Let $T : \mathscr{H} \to \mathscr{H}$ be nonexpansive with $\text{Fix}(T) \neq \emptyset$. Let $\{x_n\}$ be the sequence generated by Algorithm 9. Then, for any $x^* \in \text{Fix}(T)$, we have*

$$\mathbb{E}(\|x_{n+1} - x^*\|^2 \mid \mathscr{X}_n) \leq \|x_n - x^*\|^2 + \frac{\gamma}{m} \sum_{d \in J(n)} \|x_d - x_{d+1}\|^2$$

$$+ \frac{1}{m} \left(\frac{|J(n)|}{\gamma} + \frac{1}{mp_{\min}} - \frac{1}{\eta_n} \right) \|x_n - \bar{x}_{n+1}\|^2,$$

where $\gamma > 0$ is an arbitrary constant and $\bar{x}_{n+1} = x_n - \eta_n S(\hat{x}_n)$.

The convergence results of Algorithm 9 are given as follows.

Theorem 8.1 *Let $T : \mathscr{H} \to \mathscr{H}$ be a nonexpansive mapping with $\text{Fix}(T) \neq \emptyset$. Assume that*

$$\eta_n \in \left[\eta_{\min}, \frac{cmp_{\min}}{2\tau\sqrt{p_{\min}} + 1} \right] \quad \text{for all } \eta_{\min} > 0 \text{ and } 0 < c < 1.$$

Let $\{x_n\}$ be the sequence generated by Algorithm 9. Then, with probability one, $\{x_n\}$ converges weakly to a fixed point of T. This convergence becomes strong if $\mathscr{H}$ has a finite dimension. In addition, if T is demicompact, then, with probability one, $\{x_n\}$ converges strongly to a fixed point of T.

Note that, in Theorem 8.1, the weak convergence result only requires T to be nonexpansive and have a fixed point.

We may check the relaxation parameter bound $\frac{cmp_{\min}}{2\tau\sqrt{p_{\min}}+1}$ in Theorem 8.1. Consider the uniform case: $p_{\min} \equiv p_i \equiv \frac{1}{m}$. Then the bound simplifies to $\frac{1}{1+2\tau/\sqrt{m}}$. If the maximal delay is no more than the square root of the number of coordinates, i.e., $\tau = O(\sqrt{m})$, then the bound is $O(1)$. In general, τ depends on several factors such as problem structure, system architecture, load balance, etc. If all updates and agents are identical, then τ is proportional to p, the number of agents.

The linear convergence is established under the assumption that S is quasi-strongly monotone. Note that the corresponding operators in the class of structured minimization with a close proper convex function and a strongly convex and Lipschitz differentiable function are quasi-strongly monotone:

Theorem 8.2 *Assume that S is quasi-μ-strongly monotone with $\mu > 0$. Let $\beta \in (0, 1)$ and $\{x_n\}$ be the sequence generated by Algorithm 9 with a constant relaxation parameter $\eta \in (0, \min\{\underline{\eta}_1, \underline{\eta}_2\}]$, where*

$$\underline{\eta}_1 = \left(1 - \frac{1}{\rho}\right) \frac{m\sqrt{p_{min}}}{8} \frac{\rho^{1/2} - 1}{\rho^{(\tau+1)/2} - 1} \quad \text{for some } \rho > 1,$$

and

$$\underline{\eta}_2 = \frac{-b + \sqrt{b^2 + 4(1 - \beta)a}}{2a},$$

$$a = \frac{2\beta\mu\tau}{m^2 p_{min}} \frac{\rho(\rho^\tau - 1)}{\rho - 1}, \quad b = \frac{1}{mp_{min}} + \frac{2}{m}\sqrt{\frac{\rho(\rho^\tau - 1)\tau}{(\rho - 1)p_{min}}}.$$

Then we have

$$\mathbb{E}(\|x_n - x^*\|^2) \leq \left(1 - \frac{\beta\mu\eta}{m}\right)^n \|x_0 - x^*\|^2,$$

where x^* is a unique fixed point of T.

Assume that i_n is chosen uniformly at random, i.e., $p_{min} \equiv p_i \equiv \frac{1}{m}$. We consider the case when m and τ are large. Let $\sqrt{\rho} = 1 + \frac{1}{\tau}$. Then it is easy to verify that $\underline{\eta}_1 = O\left(\frac{\sqrt{m}}{\tau^2}\right)$ and $\underline{\eta}_2 = O\left(\frac{\sqrt{m}}{\tau}\right)$. Therefore, if $\tau = O\left(m^{\frac{1}{4}}\right)$, then the relaxation parameter in Theorem 8.2 can be $\eta = O(1)$. Hence the linear convergence rate can be achieved.

Recently, by combining asynchronous algorithms with the inertial acceleration techniques, Stathopoulos and Jones [186] proposed the inertial asynchronous and parallel Krasnosel'skiĭ–Mann iteration. Heaton and Censor [100] introduced the asynchronous sequential inertial algorithmic framework for finding a point in the intersection of fixed point sets of a finite collection of operators.

8.2 The Cyclic Coordinate Update Algorithm

Let $T : \mathcal{H} \to \mathcal{H}$ be a nonexpansive mapping and $S = \text{Id} - T$. Then the fixed point problem (8.1) is equivalent to the following problem:

$$\text{Find } x^* \in \mathcal{H} \text{ such that } S(x^*) = 0. \tag{8.6}$$

Assume that the variable x has m blocks, $x_i \in \mathcal{H}_i$ for each $i = 1, \cdots, m$, where $\mathcal{H}_i$ is a Hilbert space. Their Cartesian product is $\mathcal{H} := \mathcal{H}_1 \times \cdots \times \mathcal{H}_m$. For each coordinate $i = 1, \cdots, m$, we define the i-th coordinate operator $S_i : \mathcal{H} \to \mathcal{H}$ such that

$$S_i(x) = (0, \cdots, (S(x))_i, \cdots, 0) \quad \text{for all } x \in \mathcal{H}.$$

Since T is nonexpansive, it follows from Lemma 7.3 that S is 2-Lipschitz:

$$\|S(x) - S(y)\| \le 2\|x - y\| \quad \text{for all } x, y \in \mathscr{H}.$$

Since $\|S_i(x) - S_i(y)\| \le \|S(x) - S(y)\|$, each S_i is 2-Lipschitz, too. However, the Lipschitz constant for S_i can be (much) smaller than 2. Hence we denote by L_i the Lipschitz constant of S_i, that is,

$$\|S_i(x) - S_i(y)\| \le L_i\|x - y\| \quad \text{for all } x, y \in \mathscr{H}$$

and set

$$L := \max_{1 \le i \le m} L_i \le 2.$$

Chow et al. [39] introduced the following cyclic coordinate update:

Algorithm 10 The cyclic coordinate update

---
input: $x_0 \in \mathscr{H}$
for $n = 1, 2, \cdots,$ **do**
    set $(i_1, i_2, \cdots, i_m)$ as a permutation of $(1, 2, \cdots, m)$;
    (either no permutation, random shuffling, or greedy ordering);
    choose a relaxation parameter $\lambda_n > 0$;
    initialize $y_0 \leftarrow x_{n-1}$;
for $j = 1, 2, \cdots, m,$ **do**
    set $y_j \leftarrow y_{j-1} - \lambda_n S_{i_j}(y_{j-1})$;
    set $x_n \leftarrow y_m$.

Each n specifies an outer loop, which is also called an *epoch*, and each j is an inner loop. Each inner iteration only updates the i_j-th coordinate:

$$(y_j)_i = \begin{cases} (y_{j-1})_i - \lambda_n(S(y_{j-1}))_i, & \text{if } i = i_j, \\ (y_{j-1})_i, & \text{otherwise.} \end{cases} \quad \text{for each } i = 1, \cdots, m.$$

The order of m coordinates is specified at the beginning of each epoch. Each epoch selects the order independently. Typical choices include the natural ordering $(1, 2, \cdots, m)$, a random shuffling, and a greedy ordering (for example, place a coordinate earlier if the Lipschitz constant of the coordinate is larger). One can also shuffle only in the first epoch and then use the same ordering for all remaining epochs. In general, shuffling avoids the worst ordering and may accelerate the convergence.

In Algorithm 10, the relaxation parameter is taken as the same value λ_n each epoch. According to Lemma 7.2, the optimal choice of the relaxation parameters heavily depends on the properties of the operators. Therefore, it is better to replace λ_n with $\lambda_{n_{i_j}}$ and make the relaxation parameter change with n and i_j (see, for example, [69]).

The convergence result of Algorithm 10 is given as follows.

Theorem 8.3 *Let* $T : \mathscr{H} \to \mathscr{H}$ *be a nonexpansive mapping with* $Fix(T) \neq \emptyset$. *Assume that the relaxation parameter is defined by*

$$\lambda_n = \frac{1}{n^{1/2}}. \tag{8.7}$$

Then the sequence $\{x_n\}$ *generated by Algorithm* 10 *weakly converges to a fixed point of* T.

Recently, Peng and Xu [167] relaxed the condition (8.7) to the following:

$$(G1) \ \sum_{n=0}^{+\infty} \lambda_n = +\infty \quad \text{and} \quad (G2) \ \sum_{n=0}^{+\infty} \lambda_n^3 < +\infty. \tag{8.8}$$

It is obvious that the condition (8.8) includes (8.7) as a special case. Note that the relaxation parameters λ_n satisfying (8.7) or (8.8) are decreasing and go to 0 as $n \to \infty$. This generally results in the slow convergence of Algorithm 10.

Chow et al. [39] also presented the reduction rate of running minimal residual of Algorithm 10.

Theorem 8.4 *Under the setting of Theorem* 8.3, *the reduction rate of running minimal residual is*

$$\min_{j \leq n}\{\|x_j - T(x_j)\|^2\} = o\left(\frac{1}{\sqrt{n}}\right).$$

From Theorem 8.4, it follows that $\min_{j \leq n}\{\|x_j - T(x_j)\|\} = o(1/n^{1/4})$.

Under the assumption that S is quasi-strong monotone, Chow et al. [39] gave the linear convergence of Algorithm 10 with the fixed relaxation parameter $\lambda_n \equiv \lambda < 1$.

Theorem 8.5 *Let* $T : \mathscr{H} \to \mathscr{H}$ *be a nonexpansive mapping with* $Fix(T) \neq \emptyset$ *and* $Id - T$ *be quasi-*μ*-strong monotone. Assume that the relaxation parameter* $\lambda_n \equiv \lambda$ *for each* $n \geq 0$ *satisfies*

$$\lambda = \min\left\{\frac{1}{4mL}, \frac{\mu}{4\sqrt{2}mL}, \frac{2mL}{17mL + 2\mu^2}\right\}. \tag{8.9}$$

Let $\{x_n\}$ *be generated by Algorithm* 10. *Then it holds*

$$\|x_n - x^*\|^2 \leq \rho^n \|x_0 - x^*\|^2,$$

where $\rho = 1 - \frac{\lambda\mu^2}{2} < 1$ *and* x^* *is a unique fixed point of* T. *When* m *and* $\frac{L}{\mu}$ *are large, one has* $\lambda = O(\frac{\mu}{mL})$.

Acknowledgments The first author would like to thank Professor Yuchao Tang for the helpful discussion and valuable suggestions.

Conclusion

The Krasnosel'skiĭ–Mann iteration is the most popular algorithm to solve the fixed point problems. Its role in the construction and analysis of the iterative algorithms, especially operator splitting methods, in optimizations has recently paid increasing attentions. Relaxation and inertia are two main factors that affect the convergence speed of the Krasnosel'skiĭ–Mann iteration. Although some progress has been made on the range and selections of the inertial and relaxation parameters, there are few results on the optimal choices. Much of the space remains to be explored for the Krasnosel'skiĭ–Mann iteration.

Q.-L. Dong et al., *The Krasnosel'skiĭ-Mann Iterative Method*, SpringerBriefs in Optimization, https://doi.org/10.1007/978-3-030-91654-1

References

1. D.G.M. Anderson, *Comments on "Anderson Acceleration, Mixing and Extrapolation"*, Numer. Algor. **80**(2019), 135–234.
2. D.G.M. Anderson, *Iterative procedures for nonlinear integral equations*, J. Assoc. Comput. Mach. **12**(1965), 547–560.
3. F.J. Aragón Artacho, J.M. Borwein, *Global convergence of a non-convex Douglas-Rachford iteration*, J. Glob. Optim. **57**(2013), 753–769.
4. H. Attouch, J. Peypouquet, *The rate of convergence of Nesterov's accelerated forward-backward method is actually faster than* $1/k^2$, SIAM J. Optim. **26**(2016), 1824–1834.
5. H. Attouch, J. Peypouquet, P. Redont, *Backward-forward algorithms for structured monotone inclusions in Hilbert spaces*, J. Math. Anal. Appl. **457**(2016), 1095–1117.
6. H. Attouch, J. Peypouquet, P. Redont, *On the fast convergence of an inertial gradient-like dynamics with vanishing viscosity*, Technical Report arXiv:1507.04782, 2015.
7. J.P. Aubin, H. Frankowska, *Set-Valued Analysis*, Birkhäuser, Boston, 1990.
8. H. Avron, A. Druinsky, A. Gupta, *Revisiting asynchronous linear solvers: Provable convergence rate through randomization*, in Proc. the 28th Internat. IEEE Symposium on Parallel and Distributed Processing, IEEE Press, Piscataway, NJ, 2014, pp. 198–207.
9. J. Bahi, J.C. Miellou, K. Rhofir, *Asynchronous multisplitting methods for nonlinear fixed point problems*, Numer. Algor. **15**(1997), 315–345.
10. J.B. Baillon, R.E. Bruck, *The rate of asymptotic regularity is* $O(1/\sqrt{n})$, in Theory and Applications of Nonlinear Operators of Accretive and Monotone Types, Lecture Notes in Pure and Applied Mathematics, Vol. 178, Dekker, New York, 1996, pp. 51–81.
11. J.B. Baillon, R.E. Bruck, *Optimal rates of asymptotic regularity for averaged nonexpansive mappings*, In Fixed Point Theory and Applications (Halifax, NS, 1991), World Scientific, River Edge, NJ, 1992, pp. 27–66.
12. J.B. Baillon, R.E. Bruck, S. Reich, *On the asymptotic behavior of non-expansive mappings and semigroups in Banach spaces*, Houston J. Math. **4**(1978), 1–9.
13. J. Barzilai, J. Borwein, *Two point step gradient methods*, IMA J. Numer. Anal. **8**(1988), 141–148.
14. G.M. Baudet, *Asynchronous iterative methods for multiprocessors*, J. ACM, **25**(1978), 226–244.
15. H.H. Bauschke, P.L. Combettes, *Convex Analysis and Monotone Operator Theory in Hilbert Spaces*, Springer, 2nd ed, Berlin, Springer, 2017.
16. H.H. Bauschke, P.L. Combettes, *A weak-to-strong convergence principle for fejer-monotone methods in Hilbert spaces*, Math. Oper. Res. **26**(2001), 248–264.

17. A. Beck, M. Teboulle, *A fast iterative shrinkage-thresholding algorithm for linear inverse problems*, SIAM J. Imaging Sci. **2**(2009), 183–202.
18. V. Berinde, *Iterative approximation of fixed points*, Springer, 2006.
19. E. Blum, W. Oettli, *From optimization and variational inequalities to equilibrium problems*, The Mathematics Student **63** (1994), 123–145.
20. D. Borwein, J. Borwein, *Fixed point iterations for real functions*, J. Math. Anal. Appl. **157**(1991), 112–126.
21. J. Borwein, S. Reich, I. Shafrir, *Krasnosel'skiĭ-Mann iterations in normed spaces*, Canada. Math. Bull. **35**(1992), 21–28.
22. R.I. Bot, E.R. Csetnek, *A dynamical system associated with the fixed points set of a nonexpansive operator*, J. Dyn. Diff. Equat. **29**(2017), 155–168.
23. R.I. Bot, E.R. Csetnek, *An inertial forward-backward-forward primal-dual splitting algorithm for solving monotone inclusion problems*, Numer. Algor. **71**(2016), 519–540.
24. R.I. Bot, E.R. Csetnek, C. Hendrich, *Inertial Douglas-Rachford splitting for monotone inclusion problems*, Appl. Math. Comput. **256**(2015), 472–487.
25. R.I. Bot, E.R. Csetnek, D. Meier, *Inducing strong convergence into the asymptotic behaviour of proximal splitting algorithms in Hilbert spaces*, Optim. Methods Softw. **34**(2019), 489–514.
26. M. Bravo, R. Cominetti, *Sharp convergence rates for averaged nonexpansive maps*, Israel J. Math. **227**(2018), 163–188.
27. M. Bravo, R. Cominetti, M. Pavez-Signé, *Rates of convergence for inexact Krasnosel'skii-Mann iterations in Banach spaces*, Math. Program. (Ser. A) **175**(2019), 241–262.
28. L.M. Briceño-Arias, P.L. Combettes, *A monotone+skew splitting model for composite monotone inclusions in duality*, SIAM J. Optim. **21**(2011), 1230–1250.
29. Y. Censor, *Weak and strong superiorization: between feasibility-seeking and minimization*, An. St. Univ. Ovidius Constanta Ser. Mat. **23**(2015), 41–54.
30. Y. Censor, A. Gibali, S. Reich, *The subgradient extragradient method for solving variational inequalities in Hilbert space*, J. Optim. Theory Appl. **148**(2011), 318–335.
31. Y. Censor, R. Davidi, G.T. Herman, *Perturbation resilience and superiorization of iterative algorithms*, Inverse Probl. **26**(2010), 065008.
32. Y. Censor, A.J. Zaslavski, *Strict Fejér monotonicity by superiorization of feasibility-seeking projection methods*, J. Optim. Theory Appl. **165**(2015), 172–187.
33. V. Cevher, B.C. Vũ, A. Yurtsever, *Inertial three-operator splitting method and applications*, https://arxiv.org/abs/1904.12980.
34. A. Chambolle, Ch. Dossal, *On the convergence of the iterates of the fast iterative shrinkage/thresholding algorithm*, J. Optim. Theory. Appl. **166**(2015), 968–982.
35. A. Chambolle, T. Pock, *A first-order Primal-Dual algorithm for convex problems with applications to imaging*, J. Math. Imaging Vis. **40**(2011), 120–145.
36. D. Chazan, W. Miranker, *Chaotic relaxation*, Linear Algebra Appl. **2**(1969), 199–222.
37. C.E. Chidume, *Geometric properties of Banach spaces and nonlinear iterations*, Springer Verlag Series: Lecture Notes in Mathematics, Vol. 1965 (2009) XVII, 326 pp. ISBN 978-1-84882-189-7.
38. C.E. Chidume, S.A. Mutangadura, *An example on the Mann iteration method for Lipschitz pseudocontractions*, Proc. Amer. Math. Soc. **129**(2001), 2359–2363.
39. Y.T. Chow, T. Wu, W. Yin, *Cyclic coordinate-update algorithms for fixed-point problems: analysis and applications*, SIAM J. Sci. Comput. **39**(2017), A1280–A1300.
40. P.L. Combettes, *Inconsistent signal feasibility problems: Least-squares solutions in a product space*, IEEE T. Signal Proces. **42**(1994), 2955–2966.
41. P.L. Combettes, *Fejér monotonicity in convex optimization*, In A. Christodoulos Floudas, M. Panos Pardalos (Eds), Encyclopedia of Optimization, pp. 1016–1024. Springer, Boston, MA, 2001.
42. P.L. Combettes, *Solving monotone inclusions via compositions of nonexpansive averaged operators*, Optimization **53**(2004), 475–504.

43. P.L. Combettes, L. Condat, J.C. Pesquet, B.C. Vũ, *A Forward-Backward view of some Primal–Dual optimization methods in image recovery*, In Image Processing (ICIP), 2014 IEEE International Conference on pp. 4141–4145. IEEE, 2014.

44. P.L. Combettes, S.A. Hirstoaga, *Equilibrium programming in Hilbert spaces*, J. Nonlinear Convex Anal. **6**(2005), 117–136.

45. P.L. Combettes, L.E. Glaudin, *Quasi-nonexpansive iterations on the affine hull of orbits: from Mann's mean value algorithm to inertial methods*, SIAM J. Optim. **27**(2017), 2356–2380.

46. P.L. Combettes, T. Pennanen, *Generalized Mann iterates for constructing fixed points in Hilbert spaces*, J. Math. Anal. Appl. **275**(2002), 521–536.

47. P.L. Combettes, J.C. Pesquet, *Fixed point strategies in data science*, IEEE T. Signal Proces. **69**(2021), 3878–3905.

48. P.L. Combettes, B.C. Vũ, *Variable metric forward-backward splitting with applications to monotone inclusions in duality*, Optimization **63**(2014), 1289–1318.

49. P.L. Combettes, V.R. Wajs, *Signal recovery by proximal Forward-Backward splitting*, Multiscale Model Sim. **4**(2005), 1168–1200.

50. P.L. Combettes, I. Yamada, *Compositions and convex combinations of averaged nonexpansive operators*, J. Math. Anal. Appl. **425**(2015), 55–70.

51. R. Cominetti, J.A. Soto, J. Vaisman, *On the rate of convergence of Krasnosel'skii-Mann iterations and their connection with sums of Bernoullis*, Israel J. Math. **199**(2014), 757–772.

52. L. Condat, *A primal–dual splitting method for convex optimization involving Lipschitzian, proximable and linear composite terms*, J. Optimi. Theory Appl. **158**(2013), 460–479.

53. F. Cui, Y. Tang, Y. Yang, *An inertial three-operator splitting algorithm with applications to image inpainting*, Appl. Set-Valued Anal. Optim. **1**(2019), 113–134.

54. Y.H. Dai, L.Z. Liao, *R-linear convergence of the Barzilai and Borwein gradient method*, IMA J. Numer. Anal. **22**(2002), 1–10.

55. D. Davis, W. Yin, *A three-operator splitting scheme and its optimization applications*, Set-Valued Variat. Anal. **25**(2017), 829–858.

56. K. Deimling, *Zeros of accretive operators*, Manuscripta Math. **13**(1974), 365–374.

57. T. Dominguez Benavides, G. López Acedo, H.K. Xu, *Iterative solutions for zeros of accretive operators*, Math. Nachr. **248-249**(2003), 62–71.

58. Q.L. Dong, Y.J. Cho, Th.M. Rassias, *General inertial Mann algorithms and their convergence analysis for nonexpansive mappings*, pp. 175–191, Applications of Nonlinear Analysis, Edited by by Th.M. Rassias, Springer, 2018.

59. Q.L. Dong, Y.J. Cho, L.L. Zhong, Th.M. Rassias, *Inertial projection and contraction algorithms for variational inequalities*, J. Global Optim. **70**(2018), 687–704.

60. Q.L. Dong, A. Gibali, D. Jiang, S.H. Ke, *Convergence of projection and contraction algorithms with outer perturbations and their applications to sparse signals recovery*, J. Fixed Point Theory Appl. **20**(2018), 16.

61. Q.L. Dong, S. He, *A viscosity projection method for class T mappings*, An. Sti. U. Ovid. Co-Mat. **21**(2)(2013), 95–109.

62. Q.L. Dong, S. He, Y.J. Cho, *A new hybrid algorithm and its numerical realization for two nonexpansive mappings*, Fixed Point Theory Appl. **2015**(2015), 150.

63. Q.L. Dong, S. He, X. Liu, *Rate of convergence of Mann, Ishikawa, Noor and SP-iterations for continuous functions on an arbitrary interval*, J. Inequal. Appl. **2013**(2013), 269.

64. Q.L. Dong, S. He, F. Su, *Strong convergence theorems by shrinking projection methods for class T mappings*, Fixed Point Theory Appl. **681214**(2011) 7 pp.

65. Q.L. Dong, J. Huang, X.H. Li, Y.J. Cho, Th.M. Rassias, *MiKM: Multi–step inertial Krasnosel'skiĭ–Mann algorithm and its applications*, J. Global Optim. **73**(2019) 801–824.

66. Q.L. Dong, S.H. Ke, Y.J. Cho, Th.M. Rassias, *Convergence theorems and convergence rates for the general inertial Krasnosel'skiĭ–Mann iteration algorithm*, Advances in Metric Fixed Point Theory and Applications, Springer, 2021, Y.J. Cho, M. Jleli, M. Mursaleen, B. Samet, C.V. Vetro (Eds), pp. 61–83.

67. Q.L. Dong, X.H. Li, Y.J. Cho, Th.M. Rassias, *Multi-step inertial Krasnosel'skiĭ–Mann iteration with new inertial parameters arrays*, J. Fixed Point Theory Appl. **23**(2021), 44.

68. Q.L. Dong, X.H. Li, S. He, *Outer perturbations of a projection method and two approxima-tion methods for the split equality problem,* Optimization **67**(2018), 1429–1446.

69. Q.L. Dong, X.H. Li, Th.M. Rassias, *Two projection algorithms for a class of split feasibility problems with jointly constrained Nash equilibrium models,* Optimization **70**(2021), 871–897.

70. Q.L. Dong, Y.Y. Lu, *A new hybrid algorithm for a nonexpansive mapping,* Fixed Point Theory Appl. **2015**(2015), 37.

71. Q.L. Dong, Y.Y. Lu, J. Yang, *The extragradient algorithm with inertial effects for solving the variational inequality,* Optimization **65**(2016), 2217–2226.

72. Q.L. Dong, H.Y. Yuan, *Accelerated Mann and CQ algorithms for finding a fixed point of a nonexpansive mapping,* Fixed Point Theory Appl. **2015**(2015), 125.

73. Q.L. Dong, J. Zhao, S. He, *Bounded perturbation resilience of the viscosity algorithm,* J. Inequal. Appl. **2016**(2016), Art. ID 299.

74. Y. Dong, *New inertial factors of the Krasnosel'skiĭ-Mann iteration,* Set-Valued Variat. Anal. **29**(2021), 145–161.

75. W.G. Dotson, *On the Mann iterative process,* Trans. Amer. Math. Soc. **149**(1970), 65–73.

76. J. Douglas, H.H. Rachford, *On the numerical solution of heat conduction problems in two and three space variables,* Trans. Amer. Math. Soc. **82**(1956), 421–439.

77. D.A. Donzis, K. Aditya, *Asynchronous finite-difference schemes for partial differential equations,* J. Comput. Phys. **274**(2014), 370–392.

78. J. Eckstein, D.P. Bertsekas, *On the Douglas-Rachford splitting method and the proximal point algorithm for maximal monotone operators,* Math. Program. **55**(1992), 293–318, 1992.

79. M. Edelstein, *A remark on a theorem of M.A. Krasnosel'skiĭ,* Amer. Math. Monthly **73**(1966), 509–501.

80. K. Fan, *A minimax inequality and applications, Inequalities III,* Proc. Third Sympos., Univ. California, Los Angeles, 1969.

81. L. Fang, P.J. Antsaklis, *Information consensus of asynchronous discrete-time multiagent systems,* in Proceedings of the 2005 American Control Conference, IEEE Press, Piscataway, NJ, 2005, pp. 1883–1888.

82. R.L. Franks, P.P. Marzec, *A theorem on mean value iterations,* Proc. Amer. Math. Soc. **30**(1971), 324–326.

83. A. Fu, J. Zhang, S. Boyd, *Anderson accelerated Douglas–Rachford splitting,* SIAM J. Sci. Comput. **42**(2020), A3560–A3583.

84. E. Garduño, G.T. Herman, *Superiorization of the ML-EM algorithm,* IEEE Trans. Nucl. Sci. **61**(2014), 162–172.

85. A. Genel, J. Lindenstrauss, *An example concerning fixed points,* Israel J. Math. **22**(1975), 81–86.

86. P. Giselsson, M. Fält, S. Boyd, *Line search for averaged operator iteration,* In: 2016 IEEE 55th Conference on Decision and Control (CDC), pp. 1015–1022. IEEE, 2016.

87. K. Goebel, W.A. Kirk, *Iteration processes for nonexpansive mappings, in Topological Methods in Nonlinear Functional Analysis,* Contemporary Math. Vol. 21, Amer. Math. Soc., Providence, RI, 1983, pp. 115–123.

88. K. Goebel, S. Reich, *Uniform Convexity, Hyperbolic Geometry, and Nonexpansive Mappings,* Marcel Dekker, New York, 1984.

89. C.W. Groetsch, *A note on segmenting Mann iterates,* J. Math. Anal. Appl. *40*(1972), 369–372.

90. B. Halpern, *Fixed points of nonexpanding maps,* Bull. Amer. Math. Soc. **73**(1967), 957–961.

91. G.J. Hartman, G. Stampacchia, *On some nonlinear elliptic differential equations,* Acta Math. **112** (1966), 271–310.

92. Y. Haugazeau, *Sur les Inéquations Variationnelles et la Minimisation de Fonctionnelles Convexes,* Thèse, Université in Paris, Paris, France. (1968).

93. B.S. He, *A class of projection and contraction methods for monotone variational inequalities,* Appl. Math. Optim. **35**(1997), 69–76.

94. S. He, Q.L. Dong, H. Tian, X.H. Li, *On the optimal relaxation parameters of Krasnosel'skiĭ–Mann iteration,* Optimization, **70**(2021), 1959–1986.

95. S. He, T. Wu, Y.J. Cho, Th.M. Rassias, *Optimal parameter selections for a general Halpern iteration,* Numer. Algor. **82**(2019), 1171–1188.

96. S. He, C. Yang, *Boundary point algorithms for minimum norm fixed points of nonexpansive mappings,* Fixed Point Theory Appl. **2014**(2014), 56.

97. S. He, Z. Yang, *A modified successive projection method for Mann's iteration process,* J. Fixed Point Theory Appl. **21**(2019), 9.

98. S. He, C. Yang, P. Duan, *Realization of the hybrid method for Mann iterations,* Appl. Math. Comput. **217**(2010), 4239–4247.

99. S. He, J. Zhao, M. Li, *Degree of convergence of modified averaged iterations for fixed points problems and operator equations,* Nonlinear Anal. **71**(2009), 4098–4104.

100. H. Heaton, Y. Censor, *Asynchronous sequential inertial iterations for common fixed points problems with an application to linear systems,* J. Global Optim. **74**(2019), 95–119.

101. G.T. Herman, R. Davidi, *Image reconstruction from a small number of projections,* Inverse Probl. **24**(2008), 045011.

102. G.T. Herman, E. Garduño, R. Davidi, Y. Censor, *Superiorization: an optimization heuristic for medical physics,* Med. Phys. **39**(2012), 5532–5546.

103. M.R. Hestenes, *Multiplier and gradient methods,* J. Optim. Theory Appl. **4**(1969), 303–320.

104. T.L. Hicks, J.D. Kubicek, *On the Mann iteration process in a Hilbert spaces,* J. Math. Anal. Appl. **59**(1977), 498–504.

105. D.V. Hieu, Y.J. Cho, Y.B. Xiao, *Modified extragradient algorithms for solving equilibrium problems,* Optimization **67** (2018), 2003–2029.

106. D.V. Hieu, Y.J. Cho, Y.B. Xiao, *Modified accelerated algorithms for solving variational inequalities,* Internat. J. Comput. Math. **97**(2020), 2233–2258.

107. D.V. Hieu, P.K. Anh, L.D. Muu, *Modified hybrid projection methods for finding common solutions to variational inequality problems,* Comput. Optim. Appl. **66**(2017), 75–96.

108. X. Huang, E.K. Ryu, W. Yin, *Tight coefficients of averaged operators via scaled relative graph,* J. Math. Anal. Appl. **490**(2020), 124211.

109. H. Hundal, *An alternating projection that does not converge in norm,* Nonlinear Anal. Theory **57**(2004), 35–61.

110. D.H. Hyers, G. Isac, Th.M. Rassias, *Topics in Nonlinear Analysis and Applications,* World Scientific Publishing Company, 1997.

111. S. Ishikawa, *Fixed points and iterations of a nonexpansive mapping in a Banach space,* Proc. Amer. Math. Soc. **59**(1976), 65–71.

112. A.N. Iusem, A.R. De Pierrb, *Convergence results for an accelerated nonlinear Cimmino algorithm,* Numer. Math. **49**(1986), 367–378.

113. A.N. Iusem, B.F. Svaiter, *A variant of Korpelevich's method for variational inequalities with a new search strategy,* Optimization **42**(1997), 309–321.

114. F. Iutzeler, M.J. Hendrickx, *A generic online acceleration scheme for optimization algorithms via relaxation and inertia,* Optim. Method Softw. **34**(2019), 383–405.

115. F. Iutzeler, J. Malick, *On the proximal gradient algorithm with alternated inertia,* J. Optim. Theory Appl. **176**(2018), 688–710.

116. E.N. Khobotov, *Modification of the extragradient method for solving variational inequalities and certain optimization problems,* USSR Comput. Math. Math. Phys. **27**(1989), 120–127.

117. G.M. Korpelevich, *The extragradient method for finding saddle points and other problems,* Ekon. Mat. Metody **12**(1976), 747–756.

118. U. Kohlenbach, *A quantitative version of a theorem due to Borwein-Reich-Shafrir,* Numer. Funct. Anal. Optim. **22**(2001), 641–656.

119. U. Kohlenbach, *Uniform asymptotic regularity for Mann iterates,* J. Math. Anal. Appl. **279**(2003), 531–544.

120. M.A. Krasnosel'skiĭ, *Two remarks on the method of successive approximations* (in Russian,) Usp. Mat. Nauk. **10**(1955), 123–127.

121. W. La Cruz, *A residual algorithm for finding a fixed point of a nonexpansive mapping,* J. Fixed Point Theory Appl. **20**(3)(2018), 116.

122. W. La Cruz, *A spectral algorithm for large-scale systems of nonlinear monotone equations*, Numer. Algor. **76**(2017), 1109–1130.
123. G. Lewicki, G. Marino, *On some algorithms in Banach spaces for finding fixed points of nonlinear mappings*, Nonlinear Anal. **71**(2009), 3964–3972.
124. G. Li, T.K. Pong, *Douglas-Rachford splitting for nonconvex optimization with application to nonconvex feasibility problems*, Math. Program. **159**(2016), 371–401.
125. J. Liang, *Convergence Rates of First–Order Operator Splitting Methods*, Optimization and Control [math.OC]. Normandie Université; GREYC CNRS UMR 6072, 2016, in English.
126. J. Liang, J. Fadili, G. Peyré, *Convergence rates with inexact non–expansive operators*, Math. Program. (Ser. A) **159**(2016), 403–434.
127. J. Liang, L. Tao, C.B. Schönlieb, *Improving "Fast Iterative Shrinkage-Thresholding Algorithm": Faster, Smarter and Greedier*, arXiv preprint arXiv:1811.01430, 2018.
128. F. Lieder, *On the convergence rate of the Halpern-iteration*, Optim. Lett. **15**(2021), 405–418.
129. S.B. Lindstrom, B. Sims, *Survey: Sixty Years of Douglas–Rachford*, J. Austral. Math. Soc. **110**(2021), 333–370.
130. P.L. Lions, B. Mercier, *Splitting algorithms for the sum of two nonlinear operators*, SIAM J. Numer. Anal. **16**(1979), 964–979.
131. J. Liu, S.J. Wright, *Asynchronous stochastic coordinate descent: Parallelism and convergence properties*, SIAM J. Optim. **25**(2015), 351–376.
132. T.L. Magnanti, G. Perakis, *Solving variational inequality and fixed point problems by line searches and potential optimization*, Math. Program. **101**(2004), 435–461.
133. P.E. Mainge, *The viscosity approximation process for quasi-nonexpansive mappings in Hilbert spaces*, Comput. Math. Appl. **59**(2010), 74–79.
134. P.E. Mainge, *Convergence theorems for inertial K M-type algorithms*, J. Comput. Appl. Math. **219**(2008), 223–236.
135. P.E. Mainge, M.L. Gobinddass, *Convergence of one-step projected gradient methods for variational inequalities*, J. Optim. Theory Appl. **171** (2016) 146–168.
136. P.E. Mainge, S. Maruster, *Convergence in norm of modified Krasnosel'skiĭ–Mann iterations for fixed points of demicontractive mappings*, Appl. Math. Comput. **217**(2011), 9864–9874.
137. Y. Malitski, *Golden ratio algorithms for variational inequalities*, Math. Program. **184**(2020), 383–410.
138. Y. Malitski, *Projected reflected gradient method for variational inequalities*, SIAM J. Optim. **25** (2015), 502–520.
139. Y. Malitsky, M.K. Tam, *A forward-backward splitting method for monotone inclusions without cocoercivity*, SIAM J. Optim. **30**(2020), 1451–1472.
140. Y. Malitsky, T. Pock, *A first-order primal-dual algorithm with linesearch*, SIAM J. Optim. **28**(2018), 411–432.
141. W.R. Mann, *Mean value methods in iteration*, Proc. Amer. Math. Soc. **4**(1953), 506–510.
142. G. Marino, L. Muglia, *Boundary point method and the Mann-Dotson algorithm for non-self mappings in Banach spaces*, Milan J. Math. **85**(2019), 73–91.
143. G. Marino, H.K. Xu, *Weak and strong convergence theorems for strict pseudo-contractions in Hilbert spaces*, J. Math. Anal. Appl. **329**(2007), 336–346.
144. G. Marino, H.K. Xu, *A general iterative method for nonexpansive mappings in Hilbert spaces*, J. Math. Anal. Appl. **318**(2006), 43–52.
145. C. Martinez-Yanes, H.K. Xu, *Strong convergence of the C Q method for fixed point iteration processes*, J. Nonlinear Anal. **26**(2006), 2400–2411.
146. S. Maruster, *The solution by iteration of nonlinear equations in Hilbert spaces*, Proc. Amer. Math. Soc. **63**(1977), 69–73.
147. B. Martinet, *Brève communication régularisation d'inéquations variationnelles par approximations successives*, ESAIM: Mathematical Modelling and Numerical Analysis-Modélisation Math ématique et Analyse Numérique **4(R3)**(1970), 154–158.
148. S.Y. Matsushita, *On the convergence Rates of the Krasnosel'skii-Mann iteration*, Bull. Austral. Math. Soc. **96**(2017), 162–170.

149. H. Mazaheri, S.A.M. Mohsenalhosseini, *Approximate equilibrium problems and fixed points,* Internat. J. Anal. **2013** Art. ID 659493, pp. 4.

150. A. McLennan, *Advanced Fixed Point Theory for Economics,* Singapore, Springer, 2018.

151. G.J. Minty, *Monotone (nonlinear) operators in Hilbert space,* Duke Math. J. **29**(1962), 341–346.

152. J.J. Moreau, *Proximité et Dualité dans un espace Hilbertien,* Bulletin de la Société mathématique de France **93**(1965), 273–299.

153. A. Moudafi, *A reflected inertial Krasnoselskii-type algorithm for lipschitz pseudo-contractive mappings,* Bull. Iran Math. Soc. **44** (2018) 1109–1115.

154. A. Moudafi, *Viscosity approximation methods for fixed-points problems,* J. Math. Anal. Appl. **241**(2000), 46–55.

155. A. Moudafi, M. Oliny, *Convergence of a splitting inertial proximal method formonotone operators,* J. Comput. Appl. Math. **155** (2003), 447–454.

156. W.M. Moursi, Y. Zinchenko, *A Note on the Equivalence of Operator Splitting Methods,* In: H. Bauschke, R. Burachik, D. Luke (eds), Splitting Algorithms, Modern Operator Theory, and Applications, Springer, Cham. (2019).

157. Z. Mu, Y. Peng, *A note on the inertial proximal point method,* Stat. Optim. Inf. Comput. **3**(2015), 241–248.

158. K. Nakajo, W. Takahashi, *Strong convergence theorems for nonexpansive mappings and nonexpansive semigroups,* J. Math. Anal. Appl. **279**(2003), 372–379.

159. Y.E. Nesterov, *A method for solving the convex programming problem with convergence rate* $O(1/k^2)$, Dokl. Akad. Nauk SSSR **269**(1983), 543–547 (in Russian).

160. JJ. Nieto, H.K. Xu, *Solvability of nonlinear Volterra and Fredholm equations in weighted spaces,* Nonlinear Anal. **24**(1995), 1289–1297.

161. L. Oblomskaja, *Methods of successive approximation for linear equations in Banach spaces,* USSR Compt. Math. and Math. Phys. **8**(1968), 239–253.

162. D. O'Connor, L. Vandenberghe, *On the equivalence of the primal-dual hybrid gradient method and DouglasCRachford splitting,* Math. Program. (Ser. A) **179**(2018), 85–108.

163. Z. Opial, *Weak convergence of the sequence of successive approximations for nonexpansive mappings,* Bull. Amer. Math. Soc. **73**(1967), 591–597.

164. J.M. Ortega, W.C. Rheinboldt, *Iterative Solution of Nonlinear Equations in Several Variables,* Academic, New York, 1970.

165. W. Ouyang, Y. Peng, Y. Yao, *Anderson acceleration for nonconvex ADMM based on Douglas-Rachford splitting,* Comput. Graph Forum **39**(2020), 221–239.

166. D.W. Peaceman, H.H. Rachford, *The numerical solution of parabolic and elliptic differential equations,* J. Soc. Indust. Appl. Math. **3**(1955), 28–41.

167. B. Peng, H.K. Xu, *A cyclic coordinate-update fixed point algorithm,* Carpathian J. Math. **35**(2019), 365–370.

168. Z. Peng, Y. Xu, M. Yan, W. Yin, *On the convergence of asynchronous parallel iteration with unbounded delays,* J. Oper. Res. Soc. China **7**(2019), 5–42.

169. Z. Peng, Y. Xu, M. Yan, W. Yin, *A Rock: an algorithmic framework for asynchronous parallel coordinate updates,* SIAM J. Sci. Comput. **38**(2016), 2851–2879.

170. W. Petryshyn, *Construction of fixed points of demicompact mappings in Hilbert space,* J. Math. Anal. Appl. **14**(1966), 276–284.

171. E. Picard, *Memoire sur la theorie des equations aux derivees partielles et la methode des approximations successives,* J. Math. Pures Appl. **6**(1890), 145–210.

172. B.T. Polyak, *Some methods of speeding up the convergence of iteration methods,* U.S.S.R. Comput. Math. Math. Phys. **4**(1964), 1–17.

173. B.T. Polyak, *Introduction to Optimization,* Optimization Software Inc., Publications Division: New York, 1987.

174. C. Poon, J. Liang, *Geometry of first-order methods and adaptive acceleration,* 2020, arXiv:2003.03910v1.

175. H. Raguet, J. Fadili, G. Peyré, *A generalized Forward–Backward splitting,* SIAM J. Imaging Sci. **6**(2013) 1199–1226.

176. H. Rehman, P. Kumam, A.B. Abubakar, Y.J. Cho, *The extragradient algorithm with inertial effects extended to equilibrium problems*, Comput. Appl. Math. **39**(2020), 100.

177. S. Reich, *Weak convergence theorems for nonexpansive mappings in Banach spaces,* J. Math. Anal. Appl. **67**(1979), 274–276.

178. R.T. Rockafellar, *Monotone operators and the proximal point algorithm,* SIAM J. Control Optim. **14**(1976), 877–898.

179. H. Schäefer, *Über die Methode sukzessiver Approximationen*, Jahresbericht der Deutschen Mathematiker-Vereinigung **59**(1957), 131–140.

180. Y. Shehu, Q.L. Dong, L. Liu, *Global and linear convergence of alternated inertial methods for split feasibility problems,* RACSAM **115**(2021), 53.

181. Y. Shehu, A. Gibali, *New inertial relaxed method for solving split feasibilities,* Optim. Lett., **15**(2021), 2109–2126.

182. T. Shi, S. He, *Modified hybrid algorithms for Lipschitz quasi-pseudo-contractive mappings in Hilbert spaces,* Comput. Math. Appl. **59**(2010), 2940–2950.

183. M.V. Solodov, B.F. Svaiter, *A new projection method for variational inequality problems,* SIAM J. Control Optim. **37**(1999), 765–776.

184. J.E. Spingarn, *Partial inverse of a monotone operator*, Appl. Math. Optim. **10**(1983), 247–265.

185. G. Stampacchia, *Formes bilineaires coercivites sur les ensembles convexes,* C.R. Acad. Paris **258**(1964), 4413–4416.

186. G. Stathopoulos, C.N. Jones, *An inertial parallel and asynchronous forward-backward iteration for distributed convex optimization,* J. Optimiz. Theory Appl. **182**(2019), 1088–1119.

187. Y. Su, X. Qin, *Monotone CQ iteration processes for nonexpansive semigroups and maximal monotone operators,* Nonlinear Anal. **68**(2008), 3657–3664.

188. X.C. Tai, P. Tseng, *Convergence rate analysis of an asynchronous space decomposition method for convex minimization,* Math. Comp. **71**(2002), 1105–1135.

189. W. Takahashi, Y. Takeuchi, R. Kubota, *Strong convergence theorems by hybrid methods for families of nonexpansive mappings in Hilbert spaces,* J. Math. Anal. Appl. **341**(2008), 276–286.

190. A. Themelis, P. Patrinos, *Supermann: a superlinearly convergent algorithm for finding fixed points of nonexpansive operators,* IEEE Trans. Automat. Control 2019, 64, pp. 4875–4890.

191. M. Tian, *A general iterative algorithm for nonexpansive mappings in Hilbert spaces,* Nonlinear Anal. Theory **73**(2010), 689–694.

192. P. Tseng, *A modified forward-backward splitting method for maximal monotone mappings,* SIAM J. Control Optim. **38**(2000), 431–446.

193. B.C. Vũ, *A splitting algorithm for dual monotone inclusions involving cocoercive operators,* Advan. Comput. Math. **38**(2013), 667–681.

194. F. Wang, H.K. Xu, *Approximating curve and strong convergence of the CQ algorithm for the split feasibility problem,* J. Inequal. Appl. **2010**(2010), Art. ID 102085.

195. H.K. Xu, N. Altwaijry, S. Chebbi, *Strong convergence of Mann's iteration process in Banach spaces,* Mathematics **954**(2020), 8.

196. H.K. Xu, *The Krasnosel'skiĭ–Mann Iteration Method: Recent Progresses and Applications,* The 13th International Conference on Fixed Point Theory and Its Applications (ICFPTA2019), Xinxiang China, 13, July 2019, Lecture Note.

197. H.K. Xu, *Averaged mappings and the gradient-projection algorithm,* J. Optim. Theory Appl. **150**(2011), 360–378.

198. H.K. Xu, *The parameter selection problem for Mann's fixed point algorithm,* Taiwan. J. Math. **12**(2008), 1911–1920.

199. H.K. Xu, *Viscosity approximation methods for nonexpansive mappings,* J. Math. Anal. Appl. **298**(2004), 279–291.

200. H.K. Xu, *Another control condition in an iterative method for nonexpansive mapping,* Bull. Austral. Math. Soc. **65**(2002), 109–113.

201. J. Yang, H. Liu, *A modified projected gradient method for monotone variational inequalities,* J. Optim. Theory Appl. **179**(2018), 197–211.
202. J. Yang, H. Liu, *A self-adaptive method for pseudomonotone equilibrium problems and variational inequalities,* Comput. Optim. Appl. **75**(2020), 423–440.
203. Y. Yao, H. Zhou, Y.C. Liou, *Strong convergence of a modified Krasnoselski–Mann iterative algorithm for non-expansive mappings,* J. Appl. Math. Comput. **29**(2009), 383–389.
204. C. Zhang, Q.L. Dong, J. Chen, *Multi-step inertial proximal contraction algorithms for monotone variational inclusion problems,* Carpathian J. Math. **36(1)**(2020), 159–177.
205. C. Zhang, Y. Wang, *Proximal algorithm for solving monotone variational inclusion,* Optimization **67(8)**(2018), 1197–1209.
206. J. Zhao, S. He, *A hybrid iteration scheme for equilibrium problems and common fixed point problems of generalized quasi-φ-asymptotically nonexpansive mappings in Banach spaces,* Fixed Point Theory Appl. **2012**(2012), 33.

Printed in the United States
by Baker & Taylor Publisher Services